CAMBRIDGE LIBRARY COLLECTION

Books of enduring scholarly value

Zoology

Until the nineteenth century, the investigation of natural phenomena, plants and animals was considered either the preserve of elite scholars or a pastime for the leisured upper classes. As increasing academic rigour and systematisation was brought to the study of 'natural history', its subdisciplines were adopted into university curricula, and learned societies (such as the London Zoological Society, founded in 1826) were established to support research in these areas. These developments are reflected in the books reissued in this series, which describe the anatomy and characteristics of animals ranging from invertebrates to polar bears, fish to birds, in habitats from Arctic North America to the tropical forests of Malaysia. By the middle of the nineteenth century, this work and developments in research on fossils had resulted in the formulation of the theory of evolution.

The Art of Rearing Silk-Worms

This 1825 translation was published as part of a project to introduce the culture of silkworms into Britain and especially into Ireland, as Dandolo's original work was 'universally acknowledged to stand unrivalled, as at once combining theory with practice'. Vincenzo Dandolo (1758–1819), from a noble Venetian family, combined scientific and agricultural interests with a political outlook which led to his taking office under Napoleon, and retreating to a Lombardy estate after 1814. His interest in silkworms was part of a drive to improve the productivity and variety of farm produce; he also wrote on wool-bearing animals and viticulture. After an outline of the life-cycle and metamorphosis of caterpillars generally, Dandolo focuses on the silkworm. Its exclusive diet, and the specific techniques, buildings and equipment required to raise it commercially, are all discussed, as are the diseases to which it is prone, and the way to ensure a breeding stock.

The Art of Rearing Silk-Worms

Vincenzo Dandolo

CAMBRIDGE
UNIVERSITY PRESS

CAMBRIDGE
UNIVERSITY PRESS

University Printing House, Cambridge, CB2 8BS, United Kingdom

Cambridge University Press is part of the University of Cambridge.

It furthers the University's mission by disseminating knowledge in the pursuit of
education, learning and research at the highest international levels of excellence.

www.cambridge.org
Information on this title: www.cambridge.org/9781108082112

© in this compilation Cambridge University Press 2019

This edition first published 1825
This digitally printed version 2019

ISBN 978-1-108-08211-2 Paperback

The original edition of this book contains a number of oversize plates
which it has not been possible to reproduce to scale in this edition.
They can be found online at www.cambridge.org/9781108082112

Interior of a Silk Worm Laboratory.

Published by John Murray, London 1815.

THE ART

OF

REARING SILK-WORMS.

TRANSLATED FROM THE WORK OF

COUNT DANDOLO.

LONDON:

JOHN MURRAY, ALBEMARLE STREET.

MDCCCXXV

PREFACE.

In the early part of the present year certain individuals, after a careful investigation of the subject, came to a conclusion that the culture of the Silk-Worm might be carried on successfully in climates of as high northern latitude as that of Great Britain. These individuals having laid before several persons, distinguished alike for rank and talent, the arguments upon which their conclusions were founded, and having met with a favourable reception, presented a petition to His Majesty, praying that a charter might be granted them for the purpose of incorporating a company, with the power of raising such capital as might be deemed necessary for carrying into execution the cultivation of Silk in Great Britain, Ireland, and the colonies.

This prayer His Majesty's government were pleased to accede to; and one of the first acts of the company has been, to publish the following translation of the late Count Dandolo's Essay on the Cultivation of the Silk-Worm ; a work universally acknowledged to stand unrivalled, as at once combining theory with practice.

The mass of argument and facts which it will be found to contain, are, perhaps, amply sufficient to justify the most sanguine expectations of the success of the scheme, if judiciously prosecuted. As, however, that work was peculiarly addressed to climates wherein this culture is known to have flourished for a long period, a few pages will not be misapplied in putting the reader in possession of several other accredited facts, which bear strongly upon the merits of the plan, but which, from its nature, the following work cannot be expected to embrace.

The success of rearing Silk-Worms in any country depends on two circumstances, the healthy state of the insect

itself, and a sufficient supply of food for
its subsistence.

There is scarcely an individual into
whose hands this book may fall, who has
not, either in his own person or that of his
connexions, witnessed how successfully
these insects may be reared on a small
scale. In the hands of some this culture
has been carried to a considerable extent.
Success, therefore, might fairly be looked
for as undoubted, were it not that every
attempt hitherto made in this country on
a scale of magnitude sufficiently large for
the purposes of commerce has failed,
though it is an incontrovertible fact, that
were all the Silk annually grown with success by individuals for amusement collected, it would form no inconsiderable
quantity.

To what fatality can we attribute such
an apparent paradox, as that a few hundred worms shall succeed, while a few
thousand shall as constantly fail, but to
our total inattention to, and ignorance of,
their habits, rather than to the inaptitude
of our climate.

On reflection, we shall perceive that the power inherent in every animal, of destroying a certain portion of air, and rendering it unfit for the purposes of vitality, becomes dangerous, in proportion as the space in which the animal exists is diminished. A few hundred Silk-Worms, reared in an ordinary room of eight or ten feet square, where the air has free egress and ingress, possess not the power of vitiating the surrounding atmosphere so rapidly, as the current carries off the air so vitiated, and supplies its place with purer.

Provided, therefore, attention be paid to the temperature, and proper food administered, success under such conditions is inevitable. But if the number be increased to several thousands, the quantity of air vitiated in a given time will increase in an equal proportion, and as the current of air receives little or no accession of velocity, it will easily be conceived that unless recourse be had to artificial ventilation, these worms so reared will be *cæteris paribus*, under circumstances far less favourable than the former.

The chances of failure increase, in fact, in proportion to the number of worms reared in any one place; and though a freer ventilation be created, the cold thus necessarily produced will be, if not guarded against, productive in this climate of little less danger than the want of it. Hence the ill success that has hitherto accompanied attempts to rear the Silk-Worm on an extended scale.

In Italy the practice of distributing the eggs intended to produce a certain quantity of Silk, amongst the several tenantry of a proprietor, has greatly diminished the ill effects arising from improper ventilation; and the temperature of the climate is such as, generally speaking, to put the fear of cold out of the question. But even in that country, until within the last few years, it was a general and well-received opinion, that the mortality among a large number of worms was far greater in proportion than among a smaller quantity.

The cause of this circumstance the following work will be found to investigate

very fully; and they who follow the principles there laid down, need no longer fear
such a consequence.

It is not the province of these remarks
to investigate the mode of avoiding the
evils above described; the necessary mea-
sures are amply pointed out in the body
of the work. It will not, however, be
useless to beg those readers, who may desire a fuller insight into this interesting
subject, to bear ever in mind, that the
Silk-Worm is an exotic, to whose health
the torrid summers of Italy are no less
prejudicial, than the cold climate of the
north.

In its native regions it passes its short
existence on the branches of those trees
that furnish its food, exposed to the warm
breezes of an eastern climate, which supporting its life, at the same time carry off
those pestilential vapours, that are liable
to collect around it in situations more
confined. A system of management, therefore, which may as nearly as possible
assimilate our own to that climate, and
which may have the effect of placing, as

far as is in the power of imitation, these
insects in such a situation, must be the
most successful, whatever be the latitude
in which they happen to be.

Fortunately our hopes of success depend
not alone on either the truth or fallacy of
mere argument; we are in possession of
facts that equally corroborate our anticipa-
tions of success.

In Russia the cultivation of this branch
of commerce has been established for
several years in a latitude as far north as
54°, and with such success as to warrant
the establishment of manufactories for
working the native silk, and the hope that
a few years will render that country independent of Persia for the supply of this
valuable produce.

In Germany and Bavaria it has likewise
succeeded.

Some years back, silk was grown in
Prussia, of a quality which was considered superior to that even of Italy; a fact
that gives strength to an observation contained in a paragraph in the *Stockholme
Journall* for March, 1824, and reprinted in

the *Bulletin Universalle* of April, 1825. After
detailing the introduction of several plants
lately raised in Sweden, it adds, " That
" similar motives have instigated the en-
" couragement of the growth of silk in this
" country ; the idea, indeed, is not new,
" and experiments made long ago pre-
" sented encouraging results, though it
" appears that until the present moment
" the attempt has never been seriously
" made. Experiments made during the
" last year (1823) in Stockholm, for the
" purpose of discovering some indigenous
" tree capable of nourishing the Silk-
" Worm, have procured silk of very fine
" quality. The culture of the mulberry-
" tree is extending itself in the provinces
" and important communications on the
" most convenient mode of rearing the
" worm have been generally promulgated.
" The silk so produced in Sweden has
" confirmed, in the amplest manner, the
" remark formerly made on the superior
" fineness and solidity of silk grown in
" the north, compared with that from
" more temperate climes ; a fact that has

" received the unanimous sanction of the
" members of the Royal Society of Com-
" merce, as well as of many silk manufac-
" turers. It supports the ordinary prepa-
" ration and dye equally with the best
" Indian silk, possessing the same bril-
" liancy and the same softness. The silk
" also that has been grown for the last few
" years in Bavaria, is superior to that pro-
" duced in Italy."

It is true enough, that experiments, even within the present year, have been tried; some with utter failure, others with considerable success; but none of them in buildings properly adapted to secure equal warmth and ventilation; and it is satisfactory, from the great authority of this book, to collect that experiments so tried, and failing, must have equally failed, had they been tried in the same manner in any climate, or in any country of the world.

The second question, whether a sufficiency of food can be procured in northern climates, and with sufficient certainty, is in a great measure answered by the fore-

going statements; but every doubt on
that head must cease, when it is known
that though hitherto the white mulberry
(for to that alone can we look as a profitable
food for this animal) has not been an object of attention, yet it has been grown in
almost every part of England with success. At this moment, in the very pre-
cincts of London, a plantation of several
thousands of these trees is in existence;
some of them not less than fifteen years
of age, which, in spite of a soil, perhaps
the most uncongenial that could have been
selected, are flourishing, and have furnished leaves this present year, for the
prosecution of experiments on a scale of
some extent.

These trees have been purchased by the
British, Irish, and Colonial Silk Company,
and will, in the course of the ensuing
autumn, be removed to a better soil.

During the last century, some French
refugees, in the south of Ireland, made considerable plantations of this tree, and had
begun the cultivation of silk with every
appearance of the most perfect success;

their subsequent removal, however, caused
this important opportunity to be neglected, and though the trees are said to
have been very flourishing, they have since
been entirely cut down. Such facts, however, leave no doubt but that our climate
is perfectly congenial to the white mulberry, if cultivated with even moderate
attention.

A few words on the expedience of encouraging this new branch of agriculture.
Setting aside all that might very reason-
ably be argued from the increasing consequence of the silk trade, and the disadvantage of being wholly dependant on
foreign supplies, let us turn our attention
to Ireland, which, from many circumstances, appears peculiarly favourable to
such an undertaking.

The climate of Ireland is temperate;
the Irish are an agricultural people, and
much of the purposed plan is strictly
agricultural. One of the objections raised
to its success has been the value of labour. In Ireland labour is lower than on
the continent, consequently, were this ob-

jection ever so solid, there it cannot exist. Ireland possesses an enormous population of women and children; it is they who must perform those delicate operations of reeling, which the more clumsy hands of the other sex are incapable of performing.

The Irish proprietor must then, and does, view this attempt with the most intense interest,—must wish the fullest success to the objects of the company ; nor is it out of his power to aid them. Let each proprietor, following the example of the noble and distinguished Irish characters who patronise this company, devote some part of his estate to the cultivation of the white mulberry-tree, and let him feel assured that by so doing he is conferring on his country a source of inexhaustible future comfort and prosperity. But at the same time let not the proprietor of this country remain idle, and permit the sister isle to reap the entire benefit of so rich a mine ; let him recollect that, in whatever part of the united kingdom the cultivation of silk shall be established, that there the manufactures dependant on it must follow.

Impressed with these ideas, the British, Irish and Colonial Silk Company have already directed their attention to the formation of extensive plantations in England and Ireland; but fully aware themselves how greatly the success of an undertaking of such magnitude depends upon the concurrent efforts of the community at large, they desire to impress on all who can appreciate so desirable an event as the successful cultivation of silk in this country, how necessary to these efforts is their support; and considering that there is not any mode better calculated to further this object, than to spread widely a knowledge of the principles of an art so little understood, they have caused to be published this translation of a work by the late Count Dandolo, an Italian cultivator, patriot, and sçavant, whose meritorious efforts for the improvement of his country have raised his name to a high station in the annals of agriculture.

INCORPORATED BY CHARTER.

THE

BRITISH, IRISH, AND COLONIAL

SILK COMPANY,

CAPITAL £1,000,000—IN SHARES OF £50.

Auditors.

William Hill, Esq.   Joseph Sabine, F.R.S.
William Howard, Esq.   Edward Samuel Walker, Esq.

OFFICERS OF THE CORPORATION.

Bankers.

Messrs. Everett, Walkers, Maltby, Ellis, & Co.

Counsel.

William Harrison, Esq.
F. W. Sanders, Esq.,—Conveyancing Counsel.
George James Pennington, Esq.

Solicitors.

Messrs. Lowdham, Parke, & Freeth.

Law Agent in Dublin.

John Tew Armstrong, Esq.

Honorary Superintendent.

Arthur Aikin, Esq., S.S.A., F.L.S.

Architect and Surveyor.

Thomas Allason, Esq.

Secretary.

Peter Harriss Abbott, Esq.

TABLE OF CONTENTS.

CHAPTER VII.

CHAPTER VIII.

THE

ART OF REARING

SILK-WORMS.

CHAPTER I.

OF CATERPILLARS GENERALLY, AMONGST WHICH
IS COMPRISED THE SILK-WORM.

HE who contemplates nature finds an inexhaustible source of wonder and pleasure in considering,
among the class of insects, their forms, their colours, the different offensive and defensive wea-
pons with which they are provided, their curious
habitudes, the bond of union which is shewn in
some kinds, and the prudence and industry which
they employ, less indeed for their individual preservation than with a view to secure the perpe-
tuity of the species, while yielding to the soft and
powerful impulse of nature. But if this innumerable family of little animals furnishes ample
matter for the curious researches of the naturalist,
it affords also a subject of meditation for public

B

economy, since some kinds of these beings cause even national calamities, while on the contrary other species greatly contribute to the prosperity of states and individuals. It would be too long, and not necessary to the purpose I have in view, to enumerate here each family of insects: I shall confine myself to observe that those kinds which have wings present themselves to us under different states, in the various periods of their lives, and that the faculty of reproducing their species is reserved particularly for the last period. Thus one sees that, while the butterfly deposits its eggs, which are impregnated soon after copulation, these eggs do not immediately produce other butterflies, but, on the contrary, little animals of a long cylindrical form, composed of a certain quantity of segments or rings, having below a certain number of feet, of various forms and substances, with other particularities which I shall describe afterwards. Such, in general, are caterpillars, amongst which the silk-worm occupies the first place, and will form the subject of the following treatise.

In this first Chapter I shall make some observations,

1st, On the external and general characters of caterpillars.

2d, On the changes they undergo.

3d, On their manner of living, their growth, and preservation.

4th, On their transition from the state of caterpillar to that of apparent death, or chrysalis.

5th, On the change of the chrysalis into a perfect animal, or butterfly ; on the laying of the impregnated eggs, and on the death of the butterfly.

6th, On the method adopted by nature to destroy a great number of them, that they might not exceed the limits fixed for them, and on the means which man may employ with the same object in view.

1st. On the General and External Characters of Caterpillars.

Caterpillars have, as I have observed, a long body, more or less cylindrical, which is formed in its length of twelve membranous parallel rings, which, in the movements of the animal, mutually contract and elongate. They have uniformly a scaly head, of a substance similar to horn, provided with two very strong jaws, formed like a saw, which are moved horizontally, and not from above downwards, as among animals with red blood. Under the jaws is placed the contrivance by which each caterpillar deposits the silky substance. They have never fewer than eight feet, and never more than sixteen ; the six first, formed of a scaly substance similar to that of the head, are fixed under the three first rings, and can neither

B 2

be sensibly shortened or lengthened ; the others, whether two, four, six, eight, or ten in number, are membranous, flexible, and attached in pairs to the back part of the body, under their corresponding rings.

These last legs are those which transport the animal ; they are provided with little hooks, tolerably strong, calculated to give him support, and enable him to climb. All the hinder legs disappear, of whatever kind the caterpillar may be, when it changes into a butterfly, and there remain only the six first, which are variously modified. The anus is placed under the last ring. Caterpillars breathe by eighteen apertures, situated nine on each side of the body, through which the air passes in and out. Each of these openings is considered as the termination of a particular windpipe. A great number of caterpillars have eyes ; some of them are utterly blind, but they acquire the power of vision when they attain the state of butterflies. Having pointed out the general external characters which distinguish caterpillars from all other animals, it would seem useless to mention that some, according to their different kind, are large, middling, and small; but however they are all very large, compared with the egg from which they are, produced, or with themselves at the moment they emerge from the egg, as will be seen in Chapter VII.

Some caterpillars have a smooth skin, as the
silk-worm, others are rough, and elevated in certain parts; some have one either partially or entirely of velvet, covered with hair or bristles, of
various colours, often so beautiful, lively, and so
well shaded, that art cannot imitate them. It is
not a part of my subject to speak of the anatomi-
cal structure of these insects.

2. Of the Changes which Caterpillars undergo.

A character peculiar to caterpillars is to change
their skin at least three times before they reach
the period when they spin the silk which they contain, in order to assume the chrysalis state in the
cocoon or envelope which they have formed (§ 4.)

In the greatest number of caterpillars this
change takes place three or four times, in others
from five to nine times. These changes are called
casting of the skins, and are in fact diseases that
often destroy the lives of a great number of these
insects (§ 6.)

One single skin allotted to an animal which, in
a short space, increases its weight a thousand
times, would with difficulty have been able to
distend itself sufficiently to cover it entirely. Thus
provident nature has extended over the body of
the caterpillar the embryos of the skin of each
moulting, which supply also the hairs or bristles
with which many kinds are abundantly covered

As the animal increases more than the skin admits of distension, this falls off and is replaced by a second which is softer ; this again is detached in the same manner as the first, and is followed by a third, a fourth, and so on.

The slough of caterpillars is an entire covering applied to all the exterior parts of the animal, in which you recognise the hairs, the feet, the head, the skull, the jaws, teeth, &c.

As soon as the skin begins to pinch the cater-pillar, it immediately abstains more or less from food, which circumstance points out with certainty the disease occasioned by casting of the skin. This is the reason also why the caterpillar becomes much less as the time of changing its skin ap-proaches.

Then it throws out at different points of its body silky traces, which it attaches to neighbouring bodies, in order that, while it exerts itself, the skin which covers it may remain fixed in the spot where it is placed. When this operation is fi-nished the animal remains more or less quiet, and afterwards generally begins to move its head, turning itself about. Thus, the mask or scale which covers its snout being pushed forwards by the new skin formed underneath, is the first por-tion detached.

The mask being detached, the caterpillar pushes forwards, with effort, through the opening of the

first ring, which is smaller than those which follow ; and as it has already fastened the skin by
different threads, and particularly by the hooks of
the two processes of the anus, which it has already
attached where it is, it is not difficult to free its
two first feet, and afterwards to escape in a short
time from its envelope, by the help of its vermicular movements.

Sometimes the covering is broken, or a part remains attached to the extremity of the caterpillar,
which cannot always get rid of it. Then it swells
or enlarges in the part which is disentangled, while
the envelope compresses the rest of the body : in
such cases the animal dies, after a greater or less
exertion.

If after having eaten much food, the caterpillar,
approaching the time of casting its skin, had not
given a great extension to its body, which was
afterwards diminished in size by fasting and the
voiding of superfluous matter, it would not have
been able so easily to have freed itself from its
skin. A great number of caterpillars change their
skins entirely in a moment.

At this period of the life of the caterpillar nature brings about a favourable crisis, for there is
thrown out on the surface of its body a humour,
which being deposited between the old and new
skin, facilitates the extrication of the animal. The
surface of the animal is then moist.

The colour of caterpillars that have just cast their skin is pale, which serves to distinguish them from others. The new skin is much wrinkled, while the old one was dry and puckered, at the beginning. The fasting, exhaustion, inactivity, and disorder that accompany the casting of the skin are what are commonly called the sleep of the silk-worms. The caterpillar is much exhausted by this action, and the parts of his body which before were hard are now more flexible; they harden by the contact alone of the air. After the last casting of skin, each sort of caterpillar spins the silk, and prepares with it the cocoon or retreat, in which the changes into a chrysalis, and afterwards into a butterfly, or perfect animal, producing eggs that generate caterpillars, are to take place.

With most caterpillars the exterior castings of the skin, as well as those two which take place within the cocoon, require fifteen days ($ 4, 5,) In some species this process occupies one or two months, in others three months, and in some again it requires one, three, or even four years.

3. Of the Manner in which Caterpillars live, nourish, and preserve themselves.

Without speaking here of some kinds of caterpillars that live by devouring one another, all the others, which are very numerous, exist on vegetable substances, most of which are useful to man.

We often have the mortification to witness the destruction of fruit trees, of kitchen or pleasure gardens, hedges, woods, &c. &c. Often also these insects, either not finding leaves, or preferring other things to them, destroy buds and flowers, attack fruits, lodge in them, spoil them, and cause them to fall off. Many kinds of caterpillars insinuate themselves into the earth, attack the roots of herbaceous plants, shrubs, or great trees, and make them unhealthy; to the no small astonishment of the agriculturist, who sees his trees decay without any apparent cause.

Some caterpillars live in the trunks of trees, wound them, piercing them in a thousand ways, and causing them to die prematurely.

We find also frequently swarms of almost invisible caterpillars in corn, of which we become aware only when we see gnats in the granaries, or the grain itself rendered more or less light, according as the insect has more or less eaten it.

Sometimes we find the grain quite empty, though of a good appearance, because the caterpillar, which has entered by a small hole, eats only the flour, leaving the husk untouched. In a single grain, the caterpillar finds sufficient nourishment to support all its changes; which leads one to suppose that when we find grains, the flour of which is not entirely consumed, there is lodged within a caterpillar, which is not yet transformed into a perfect animal.

Every kind of caterpillar does not eat during the day; a great number of them feed only in the morning and evening: some only during the night: their voracity, however, is so great, that in a short time they destroy a quantity of food, a hundred, and even a thousand times greater than their own weight. (Chap. XIV.)

In general each kind of caterpillar lives upon a particular kind of plant; still, notwithstanding, they devour many plants, because the taste of almost every species is different: happily for us, the leaf of the mulberry is confined to the silk-worm.

Not only is there a sort of aversion for cater-pillars in general, but we think some of them are venemous, which however is not the case. The itching and slight inflammation produced by the touch of a caterpillar that has a velvety coat, is considered as a proof of venom; but nettles pro-duce the same effect.

Nature has appointed four methods for the pre-servation of all kinds of caterpillars during the severity of winter, in which they would all probably die, if they were not directed by their instinct.

Many sorts preserve their embryo during winter,

deposited in an egg, in order that at the time of
their developement in fine weather, they may immediately find proper nourishment at hand: other
kinds, born in autumn, and consequently small,
envelop themselves in many leaves, or other vegetable substances, which they unite, arrange and
attach with great skill, by means of thin silken
threads ; here they live in a torpid state, to come
forth when the leaf unfolds itself: we see also
some species preserve themselves under the form
of *nympha*, or chrysalis, in various kinds of cocoons or retreats: others again lodge themselves in
the ground, on walls, in the trunks of trees, under
stones, &c., waiting for the mild and warm season.

During winter, we meet everywhere with eggs
or caterpillars, or chrysalides, attached or suspended from trees, or fixed to walls, in the fields,
meadows, woods, &c., and always as much as
possible sheltered from cold and other inclemen-
cies, of the season. We often see also in the
winter, in the meadows, marks which point out
the nest of some kind of caterpillar.

As soon as fine weather appears, the caterpillar
emerges suddenly from the egg, develops itself,
changes into a chrysalis in the cocoon, and at
last appears in the form of a butterfly.

Among the different kinds of caterpillars, there
are some which live entirely apart ; others which
live together till the first casting of their skin, or

even longer ; while there are some which are always congregated till their change into the chrysalis.   When turned into perfect and winged
animals, they fly about in every direction.

4. *On the Transition from the Condition of the
Caterpillar, to that of apparent Death, or, to the
Chrysalis state.*

The last casting of the skin, which is visible,
being finished, the caterpillar devours, during a
certain number of days, an almost incredible quantity of food, and attains its greatest size.   Arrived at this point, its appetite abates, and entirely
ceases.   The animal then loses by degrees its
weight, and size.   (Chap. VII.)

Disgusted with its food, it seeks change of place
and solitary rest; it feels the want of clearing itself
of all the excrementitious matter which loads its organs, and even of the membrane which enveloping the excrementitious matter, served as a species
of lining to the stomach and intestines.   The silky
substance, and the animal substance, with more
or less liquid, is then all that remains of the
caterpillar.

The insect, when in this state, continues to contract its skin, and it is this contraction which
powerfully assists it in spinning, or facilitates its
means of emitting the silk, contained in its small
reservoirs.

It is evident that by this contraction, by the emission of the silk, and continual evaporation, (Chap. VIII. § 7), the skin shrinks and wrinkles, the rings approximate, and the insect becomes gradually smaller.

The formation of the chrysalis then begins, which is accomplished when the silk is entirely drawn out, and when the wrinkled slough of the animal is cast in the cocoon.

It appears from this, that nature in the various modifications of the caterpillar, only tends to the simplification of the animal.

Thus the caterpillar is in the first instance composed of animal, silky, and excremental particles: this forms the state of the *growing caterpillar;* in the next instance, it is composed of animal and silky particles; it is then the *mature caterpillar;* and, lastly, it is reduced to the animal particles alone, and is termed, in this state, the chrysalis; the cocoons, or shells, which caterpillars form, vary much in shape, as we may observe on several occasions.

Some species of caterpillars, filling small spaces begin their cocoon with several ends of silken thread, placing themselves in the centre, without however hiding themselves. Others unite and fasten together with the silky substance, one or more leaves; thus sheltered to enable themselves to draw their silk in safety previous to their transition. Others, placing the threads between the twigs or small branches of trees, drawing their silk, strive with industry to shelter the chrysalis. It is admirable to observe the art which the various species of caterpillars employ to form the silken habitation that is to harbour the chrysalis.

This art is the more ingenious, because some of the species have less silk than the others. Some are obliged to penetrate into the earth to screen their retreats. Others, to unite with their silk various substances, such as hair, and bits of leaves, &c. &c. In short, this little animal does all it can to build itself a retreat which may protect it from the rigours of the seasons, at the moment of its change to the state of chrysalis.

The period caterpillars employ in the spinning of their silk, and consequently in the construction of their habitation, varies. Some there are that construct them in an hour, others in a day, others in two days; and the silk-worm generally takes three days to spin the cocoon.

The quantity of silk which caterpillars commonly produce, is not always in proportion to their size. The silk-worm, which is not of the largest species, produces more silk than any of the others. (Chap. XIV. § 5.)

There is great variety in the colours of caterpillars; we see them of yellow, white, red, brown, sky-blue, greenish colours, and various other shades.

There is a great difference between the silk drawn by the silk-worm and that of other caterpillars. With tepid or hot water, the gummy substance, which unites the threads of the cocoons of the silk-worm, may be dissolved, which renders the silk easy to wind; whilst the threads of the silk produced by other caterpillars, adhere so closely together, there are no means of loosening and separating them; the only manner of making any use of it, would be to tear or cut the cocoons, and then card them, as is done with the *floss* or coarse interior part of the cocoon of the silk-worm; in this way, the silk of many caterpillars might become of some advantage in domestic economy, as the cocoons of common caterpillars are to be found in all places.

5th. Transition of the Chrysalis to the state of perfect Animal. or Moth.—Production of the impregnated Eggs.—Death of the Moth.

The caterpillar is subject in the course of its existence to three changes of organization.

The first change takes place when the embryo passes to the state of caterpillar, during which passage, those moultings or castings of the skin occur, which I have before noted, (Chap. I. § 11.)

The second change is the transition of the caterpillar to the state of chrysalis, or aurelia, in the cocoon it has formed for itself, (Chap. I. § 4.)

Lastly, the third change is that of the chrysalis into the perfect animal or moth, which also takes place in the cocoon.

The caterpillar, in that latter change, not only attains the state of perfect animal, but in the female the formation of the eggs takes place, and in the male the impregnating liquid is secreted.

The change of the aurelia into the moth takes place, as I before stated, in a covering enclosed within the cocoon ; the aurelia, on being entirely converted into the moth, rends this covering as well as the cocoon, and leaves the various envelopes which confined it.

The moths then couple, and the female deposits her eggs in the spot where they are best sheltered from cold, rain, or other accidents.

The term of existence in moths varies; there are some who live through the winter, and do not deposit their eggs till spring.

The males and females die very shortly after the deposition of the impregnated eggs.

6th. *Of the Means used by Nature to destroy Caterpillars, thus preventing their unbounded Increase ; means which may also be used by Man to diminish their Numbers.*

The labourer, a constant spectator of the havoc made by caterpillars, cannot imagine that provident nature should have bestowed such a curse on

mankind. He would not allow the existence of
any caterpillar, except the silk-worm, and a few of
those beautiful and bright-winged butterflies,
which ornament our museums and delight the
naturalist.

The labourer dwells little on final causes; he
imagines nature must act for him alone; he is
ignorant of the harmony and variety which exist
in the infinity of objects that compose the uni-
verse.

But although caterpillars do much mischief,
nature having assigned them a specific employment, has prescribed limits to their increase
that they might not injure us too deeply. An
innumerable quantity of these insects are naturally destroyed every year in three ways, besides
the numbers which men may destroy by artificial
means.

1st. Numerous tribes of birds feed upon the
caterpillars they find on trees or on the ground,
either when just hatched from the eggs, or when
they are grown larger.

This food, which birds particularly like, and
with which they rear their young, attracts them
to the woods, the fields, and gardens, and they
seek, with peculiar avidity, for the eggs of caterpillars.

2d. Among the various species of caterpillars,
there are some that feed upon each other, with-

out mentioning the numbers that are destroyed by lizards, frogs, toads, wasps, flies, spiders, ants, beetles and other insects, either devouring the caterpillar whole, tearing them to pieces, gnawing them, or sucking their blood. In vain does the caterpillar endeavour to avoid its innumerable foes, suspending itself by its silken thread and swinging in mid air, it scarcely ever escapes destruction.

To this species of destruction may be added that which assails some species of caterpillars by means of other insects that deposit their eggs : thus they become the prey of those insects the moment the eggs are hatched.

3d. The winter frosts and the cold spring rains kill a great number of caterpillars. We frequently observe, in spring, during the course of a few hours, these insects dropping off, struck by the shower, which chills them before they can find shelter, or at the period when they are casting their skins.

A vast number may also be destroyed in an artificial manner, by taking the nests ; those species most noxious are not in very great numbers, and are constantly within reach.

It is easy in winter to distinguish caterpillars' nests ; they are often found suspended towards the ends of boughs and branches of trees, rolled up in leaves. By the aid of large scissors (Fig. 1.)

adapted to that purpose, fixed upon a rod, and drawn by small cord, which gives action to the instrument; these boughs may be cut off with great ease.

Thus each nest destroyed, diminishes by two, or three, or four hundred, the quantity of caterpillars that would have attacked the tree.

Those who have not made use of these means which, in winter, are easy, should at least do so immediately after the spring showers, as the young caterpillars that have escaped the rain, retire into their nests, which may be easily discovered when some green or dried leaves are observed fastened and stuck together with the silky down of the caterpillar.

Those who have hedges near their gardens, or enclosures, should clip them in the winter, to prevent the caterpillars that harbour in them from falling upon their fruit-trees.

When the caterpillars are grown large, and are scattered about the shrubs, it is most difficult to prevent their ravages. They can be stupified by fumigation, and thus be made to drop off. This may be done by burning wet straw in an iron caldron or brass pan, with a long handle, (Fig. 2.) and adding a little brimstone, putting the fire at the necessary height that the smoke may reach the nests. When the smoke penetrates the parts where the caterpillars hang, the tree must be shaken, the caterpillars that drop off must be collected and burnt : but there is nothing so efficacious as destroying them in the winter.

The cabbage caterpillars may be taken at night with artificial light.

I have, doubtless, in this chapter said more than the subject of this work allows, but I thought it might be useful to avail myself of this opportunity of making a class of insects, so much complained of, more generally known, in which class·is ranked the silk-worm, that, in opposition to its species, is one of the principal sources of our wealth. What I have said may also prevent some repetitions in the course of this work.

CHAPTER II.

OF THE SILK-WORM.

WE have seen (Chap. I.) that, notwithstanding the war waged against caterpillars by men, by animals, and by seasons, in our climate ; they yet elude that destruction which threatens them so often.

It is not thus with the silk-worm. In these climates, (France and Italy), it could not thrive, nay, it could not exist through a season even, if

every care was not taken which its developement, its growth, and its perfection require : evidently proving that this insect is a native of
much warmer climates than ours.

And such, in fact, is the climate of the southern
part of the Chinese Empire, whence originated
the silk-worm, and where written documents are
preserved, tending to prove that these insects
were raised there, 2700 years before the Chris-
tian æra.

The silk-worms passed insensibly into Persia,
into India, and into various parts of Asia; they
were then conveyed to the Isle of Cos ; and in the
sixth century they were introduced into Con-
stantinople, where the emperor Justinian made
them an object of public utility. They were
successively cultivated in Greece, in Arabia, in
Spain, in Italy, in France, and in all places
where any hope could be indulged of their suc-
ceeding.

The silk-worm, being thus subjected to domestic care and habits, must, necessarily, like every
other domesticated animal, have undergone particular modifications, which have produced new
breeds or varieties, more or less different.

It is thus we may account for some silk-worms
moulting or casting their skins four times, others
only three times ; thus some form large cocoons,
nearly thrice the weight of the common cocoon.

I shall mention these breeds and varieties in the
due course of this work. (Chap. XI.)

I must here observe, that although the silk-
worm is with us. in a very different climate from
that in which it originated, and though it is domesticated, yet that we have clear proof, (Chap. V.
VI. VII.,) its constitution is sound and vigorous,
capable of often resisting the severe trials made
on it by error and ignorance. Still we sometimes
see whole broods of these insects fail in a short
period, and others lose much in quantity and
quality by want of care. (Chap. XII.) It has been
said, that in Asia they obtain as many as twelve
crops of cocoons in a year. And some person made
experiments, and published some views on the
subject, stating, that with us there might, at
least, be gathered two crops. My experiments on
the contrary tend to prove, that it would be a
sure method of destroying the mulberry-trees,
and consequently the entire breed of silk-worms.
I cannot, indeed, bring myself strictly to believe,
that in Southern Asia it should be possible to
have such numerous crops of cocoons.

The southern part of China corresponds very
nearly, as to climate, with the southern parts of
Persia; and yet there, according to the illustrious
Pallas, the boughs of the mulberry-trees are cut
only twice in the year, so as to have in the same
year two crops of cocoons.

They have admitted in Persia, upon an economical principle, the custom of feeding the silkworm upon the boughs of the mulberry-tree, and not upon the leaf alone, as we practise it, and as it is the custom in all the temperate regions. In this manner the leaves adhering to the branch are fresher and have a better flavour, and are, consequently better calculated for nutrition; the silk-worm thus eating them entirely, by these means there is no waste. In those climates they prune the small branches twice a year, because the summer being longer, the mulberry-tree is more vigorous there than it is with us.

Here, on the contrary, the mulberry-tree cannot even bear the stripping of its foliage once a year without being injured, and certainly would die if stripped twice.

All things considered, I am well persuaded that one of our good crops will be equal in produce to any crops that may be gathered elsewhere in a year.

The cocoon of the silk-worm is commonly white, straw colour, or deep yellow; we see but few here of a greenish hue, or, indeed, of any other colour. The black, and the tiger-spotted silk-worm, in general, produce cocoons of the same colour as other silk-worms.

The silk-worm is of the class of caterpillars that have the greatest number of legs; it has sixteen, that is to say, six shelly or scaly legs, and ten

membranaceous, (Chap. I.); like all other caterpillars it has neither red nor warm blood, and,
consequently, its warmth is always equal to the
temperature of the atmosphere in which it lives: it
has eighteen organs of respiration, as I stated above.
Many wrinkles are perceptible behind its
head, a small horn is placed on the last ring, at
the other extremity, and it has two reservoirs'of
silk, which unite in one aperture, through which
it draws the thread; and the colour of these reservoirs is of rather a dirty white, particularly as
they increase.

A very useful peculiarity in the silk-worm is,
that it never quits the leaf, nor the skeleton of
the leaf upon which it has been deposited, even
when hungry. The insect only rambles at the moment of its birth, before it has got the mulberry-
leaf to feed on; and when it has done feeding, and
is mature, and feels the wish of beginning its cocoon; or, lastly, when it is diseased. Excepting
in these three cases, it is never seen to wander,
but feeds on steadily without even going from
one end of the table, or tray, to the other.
Sometimes several silk-worms fasten to the inside edge of the tray, and even on the rim if they
are hungry; but as soon as they perceive the
smell of the leaves they go down. These movements commonly occur in the first stages of their
existence. It might be said with truth, that ex-

cepting these cases, there are few silk-worms that have in the course of their lives travelled beyond the space of three feet.

The time employed by the silk-worm in our climates, from the period of hatching until it deposits the seed and dies, is about sixty days. The greater the degree of heat in which it is reared, the more acute are its wants, the more rapid its pleasure, and the shorter its existence.

From the moment of the birth of the silk-worm until the gathering of the cocoons, I reckon, in my establishment, about forty days: Sometimes an unfavourable season obliges us to prolong the existence of the insects for some days, as will be seen hereafter, to enable them to digest and properly work the food necessary for them.

If the artificial heat were great and uninterrupted, the cocoons might be gathered in less than five-and-thirty days. Great heat shortens the time employed in rearing silk-worms, but it may also easily become their bane, if it be not accompanied by most constant and attentive care.

Chapter III.

OF THE ONLY FOOD PROPER FOR THE SILK-WORM.

HAVING given a general idea of the class of caterpillars, and of the species termed silk-worm, I think it may be useful, before I treat of the manner of rearing these insects, to mention their food, its different qualities, and its action upon the body of these small animals.

Whatever authors may have stated at various periods, it is demonstrated that the only leaf which agrees with the silk-worm is that of the red or white mulberry-tree

The first silk-worms reared in Europe were fed on the black mulberry leaves, the only species then cultivated among us, as would appear, although it was well known that the white mulberry was cultivated in Greece.

The white mulberry was then without delay introduced into all the temperate regions of Europe.

This mulberry-tree offers three advantages : the leaf is earlier, and thus the care of the silk-worm is not prolonged too much into the hot season. It also gives a much greater abundance of leaves in a shorter period, and the quality of the leaf produces that sort of silk most approved of by the manufacturers, although we shall see in the course of this work (Chap. VIII. § 6.) that the quality of the silk does not solely depend on the food, but also on the degree of temperature in which the silk-worm has been reared.

As there are mulberry-trees of different qualities, it might be imagined that these differences influenced the state of the silk-worms. There are five different substances in the mulberry: 1st. The solid, or fibrous substance. 2d. The colouring matter. 3d. Water. 4th. The saccharine substance. 5th. And the resinous substance.

The fibrous substance, the colouring matter, and the water, excepting that which in part composes the body of the silk-worm, cannot be said to be nutritive to the silk-worm.

The saccharine matter is that which nourishes the insect, that enlarges it, and forms its animal substance.

The resinous substance is that which, separating itself gradually from the leaf, and attracted by the animal organization, accumulates, clears itself, and insensibly fills the two reservoirs, or silk vessels, which form the integral parts of the silk-worm. According to the different proportions of the elements which compose the leaf, it follows that cases may occur in which a greater weight of leaf may yield less that is useful to the silk-worm, as well for its nourishment, as with respect to the quantity of silk obtained from the animal.

Thus the leaf of the black mulberry, hard,

harsh, and tough, which is given to silk-worms
in some of the warm climates of Europe, in Spain,
in Sicily, in Calabria, and in some parts of Greece,
&c. &c. produces abundant silk, the thread of
which is very strong, but coarse.

The white mulberry leaf of the tree planted
in high lands, exposed to cold dry winds, and in
light soil, produces generally a large quantity of
strong silk, of the purest and finest quality.

The leaf of the same tree, planted in damp situations, in low grounds, and in a stiff soil, produces less silk, and of a quality less pure and fine.

These are the most observable differences; there
are others relative to the topographical situations
of the establishments.

The less nutritive substance the leaf contains,
the more leaves must the silk-worm consume to
complete its developement.

The result must therefore be, that the silk-worm
which consumes a large quantity of leaves that
are not nutritive, must be more fatigued and more
liable to disease, than the silk-worm that eats a
smaller proportion of more nutritive leaves.

The same may be said of those leaves which,
containing a sufficiency of nutritive matter, contain little resinous substance ; in that case the
insects would thrive and grow, but probably
would not produce either a thick or strong cocoon, proportionate to the weight of the silk-

worm, as sometimes occurs in unfavourable seasons.

Notwithstanding all this, my experiments prove in the ultimate analysis, that, all things balanced, the qualities of the soil produce but a very slight difference on the quality of the leaf; that which will appear most evident is, that the principal influential cause of the fineness of the silk is the degree of temperature in which the silk-worm is reared. I have already stated this, and I shall endeavour to demonstrate it in the course of the work.

It is necessary to note not only the difference in the quality which exists in general, between the leaves of the mulberry proceeding from different soils, and picked at different seasons, but also the difference existing in the various sorts of mulberry-trees proceeding from the same soil.

I have experienced that, in equal proportion, the leaves of the broad-leaved mulberry are rather less nutritious.

I have observed that, after this species, the next is the mulberry tree that has a middle-sized leaf, thick, and of a dark-green colour. When these mulberry trees are not exposed to a dry atmosphere, and are not in light soils, they bear a large quantity of leaves, but they are not then of such a nature as to afford much silk. It appears demonstrated, that nature more easily produces leaves

abounding in nutritive matter, than those producing resinous or silky substances.

I have found that the best mulberry leaf of any
species, is that which is called the double leaf; it
is small, not very succulent, of a dark green, shining, and contains little water, which may be easily
ascertained in drying some of them ; the tree produces them in great abundance*.   In general the

* It may be of some use to subjoin a list of the varieties of
the mulberry tree hitherto described, and at the end of the
note will be found a calculation of the diminution in weight,
by drying, of each quality of leaf which I employ for the
ood of silk-worms.

1st Species. *Morus Alba.*—This species comprises the
common wild mulberry, which has four varieties in the
fruit, two have white berries, one red, and the other black ;
there are two varieties in the leaf, the one leaf deeply indented, like the leaf of the white hazel ; the other larger, not
much indented, or lobed.   The common grafted mulberry is
a variety of the first of these two that I have described, and
itself comprises the following varieties :—

1st, Of a white berry.   2d, Of a red berry.   3d, Of a
black berry.   4th, Of a large leaf, called of Tuscany.   5th,
Of a middle-sized leaf, dark-green, called in Italy *foglia gia::-
zola.*   6th, Small leaf, of a dark colour, rather thick, called
double leaf, more difficult to pick, and the best calculated
for the nutrition of silk-worms.   There are besides the fol-
owing species :—

1st, Morus tactaria.   2d, Constantinopolitana.   3d, Nigra,
(the common mulberry, well known, a fruit of a sweet flavour, particularly cultivated in the Ex-Venetian provinces.)
4th, Rubra (cultivated in botanical gardens.)   5th, Indica
(also cultivated for botanical purposes.)   6th, Latifolio (cultivated in botanical hot-houses.)   7th, Australis.   8th, La-
tifolia.   9th, Mauritiana (these three latter sort are little

cultivator has sought those mulberry trees only
which produce the largest leaves, and those of most

known in Italy.) 10th, Morus tinctoria. 11th, Morus pa-
pyrifera (these two latter species have been recently transported into another class of plants, and termed Broussone-
tia, from the name of M. Augustus Broussonet, a distinguished professor.) The list I have given sufficiently shews
the varieties of mulberry leaves which might be found even
more suitable to the nourishment of the silk-worm, than those
hitherto used.

The difference existing among the leaves of the grafted
plants is much less perceptible than the difference in the varieties of wild leaves. Thus, a wild mulberry tree of ten
years' growth, of the broad-leaved species, *unindented*, will
produce a greater weight of leaves than five trees of the same
age, of the much *indented* leaf species.

The following is the result of my experiments upon the
leaves of the grafted mulberry tree :

1. One hundred ounces of leaves nearly ripe, picked in the
same day from a Tuscany mulberry tree, produced thirty
ounces, after dessication.

2. One hundred ounces of the leaves of the giazzola mulberry tree produced thirty-one ounces and a half.

3. One hundred ounces of the double-leaved mulberry tree
produced thirty-six ounces.

This variety of the species produces more fruit than any of
the others.

All these leaves diminish still less in their weight, when they
are perfectly mature. There are few ripe leaves of different
trees which contain so little liquid as those of the mulberry
tree when ripe ; while, on the contrary, the young leaf of the
mulberry contains much liquid.

A hundred ounces of the young leaves, such as are given
to the silk-worm, in the first stage, weigh less than twenty-
one ounces when dried ; thus, it is evident, they contain almost four-fifths of water. This abundance of liquid pro-

weight, without considering that it is neither the water nor the fibre of the leaf that nourishes the silk-worm, and renders the cocoon heavy, but the resinous and saccharine substances which I have mentioned in the foregoing pages.

There is another fact to be observed; that, in equal circumstances, an old mulberry tree will always produce better leaves than the young tree; and moreover, as the tree grows older, of whatever quality it may be, the leaf diminishes in size, and improves so materially, that it at last attains a very excellent quality. I have hitherto only treated of the leaves of the grafted mulberry; the leaf of the wild mulberry is that which, in equal weight, contains, in the greatest proportion, both the nutritive and silky substances. This leaf, in a smaller quantity than that of the grafted mulberry leaf, offers a much more satisfactory result. I know not whether any one has made, with exactness, and upon a large scale, this important comparison. (Chap. XI.)

Another view of the subject, which must not be overlooked by the cultivator, is, that the grafted mulberry, particularly when old, produces a much greater quantity of the fruit than the wild tree.

This fruit, which the silk-worm generally does

vides for the very great evaporation which takes place in the body of the young silk-worm, in the first and second stage, for reasons which I have detailed elsewhere.

not eat, however, forms part of the weight of the
leaves which the cultivator either buys or sells.
These are strong reasons against the general practice of feeding silk-worms on the leaf of the wild
mulberry. (Chap. XI.)

The worst leaf that can be given to the silk-
worm, and which always injures it, is that which
is covered with what is termed *manna*, which accident arises from the diseased state of the tree, or
from its superabundant health. I would never
advise such leaves to be given to the silk-worm,
except in case of a scarcity, and then even they
should be washed and dried with great care.

The blighted, or rust-spotted leaves, do not
injure the silk-worm. We see numbers of trees
thus diseased, particularly when they are in damp
and close situations. The worm will eat this leaf,
at least will eat the sound part of it, carefully
avoiding the spots; where there is no alternative,
these leaves may be given, but in greater quantity, that the worm may not fatigue itself in seeking its necessary proportion of food. These insects would be injured by eating leaves moist with
either rain or dew. I shall shew (Chap. VII.)
how this may be avoided by drying the leaves.
The greatest care should be taken, whatever may
be the quality of the leaf, to prevent its being
heated or fermented, whether just picked or when
kept. Fermentation alters more or less the nutri-

tive substance of the leaf, which becomes less nourishing. The leaves ought not to remain long compressed in the baskets or sacks in which they are gathered.

The leaves may with ease be kept two or three days in cool, moist, sheltered places, such as cellars, storehouses, ground-floor rooms, &c., being careful not to heap them up too much, and now and then turning them to air them. They would loose their freshness in too dry a place, and might rot in too damp a situation; it is of advantage to have a place calculated for their preservation two days, or even for three days, if necessary.

The mulberry tree thrives in colder countries than Lombardy; but it should only be stripped once a year, and that crop should be gathered, so as to allow time for the leaves to shoot again before the cold weather, otherwise the tree would shortly die.

Chapter IV.

OF THE CARE NECESSARY, PREVIOUS TO THE HATCHING OF THE SILK-WORM.

THE first thing necessary to commence the operations of the year, is to detach the eggs of the silkworm from the cloths upon which they were deposited, and to prepare them for hatching.

It will be seen that this operation requires much care and attention ; but as the object is to hatch the eggs of the silk-worm at the most favourable time to ensure success, this operation may be considered as most essential.

Those who among us have reared the silk-worm, seem to have found, and still to find, it difficult to bear in mind the difference between our climate and the native warm climate in which the silk-worm originated. Obliged to have recourse to art to supply the deficiency of our climates, and to equal the advantages of those that are warmer, we have found the necessity of fixing upon some unerring method of timing the birth and rearing of the silk-worm, so as to be enabled to preserve them in health and vigour at the period most suited to our means and interest.

What has been done towards this? In past times the cultivators of silk-worms imagined that the silk-worm might be hatched at random, and spontaneously; and that if it were necessary to make an artificial climate, it was enough to use the heat of manure, or beds, or the natural heat of the body, or the kitchen fire, &c., and similar means.

It is now allowed that these methods, at best uncertain, are often pernicious to the insect. Experience having shewn this, the result has been general discouragement ; we must not therefore be surprised to observe clear proofs of the destruction of mulberry trees, and that the cultivation of them should have been given up; nor that we should see that latterly many cultivators of the silk-worm have totally despaired of rearing them with advantage.

However, since luxury has invented the hot-house, to enable us, by an artificial atmosphere, to raise exotics, it was surely natural to apply this invention to the improvement of the cultivation of silk-worms. And yet it is but very lately that this application of the invention has been thought of, which enables us in a few days to hatch with ease and certainty any given quantity of silk-worms, and rear them favourably, offering important advantages. Notwithstanding these evident facts, the art of forming these hot-houses, or stove-rooms, has not become as general as it ought to be; and farther, those who have adopted the method have not learnt the manner of conducting it with that exactness which is necessary to its success, and the advantage for which it was invented. Thus by injudicious management, whole broods of the silk-worm have been destroyed, or materially injured. I propose stating, in this chapter, the care which the eggs require to prepare them for the favourable developement of the worm, and the care necessary to fix and continue the requisite degree of temperature. We shall thus treat of,

1st, The preliminary preparation of the eggs.

2d, Of the necessity of fixing, by the thermometer, the temperature calculated to favour the hatching of the egg, and the rearing of the silk-worm.

3d, Of the hot-house, or stove-room, in which
they must be hatched.

4th, Of the hatching.

1. We suppose the eggs to be good and well
preserved, as I shall indicate in the course of this
work. (Chap. X.)

Towards the end of March, the cloths upon
which the eggs are fastened are conveyed into a
room calculated for the purpose; the cloths,
being doubled, should be put into a pail of well
or cistern-water, steeped up and down, that they
may be thoroughly soaked, and they should lie
in the water nearly six minutes, which will be
sufficient to dissolve the gummy substance by
which the eggs are stuck to the cloth. There
must be in this room tables proportioned to
the size of the cloths. The six minutes elapsed,
the cloths must be taken out, and the water allowed to drip from them, by holding them up for
two or three minutes. They should then be placed
upon the tables, partially or entirely spread out.

The cloth should be kept well stretched while
the eggs are separated from the cloth; with a
scraper, (Fig. 3.) they come off by degrees: the
scraper should not be too sharp, for fear of cutting

the eggs; neither too blunt, lest it should crush them. The eggs do not stick fast on wet linen.

When a good quantity of the eggs have been scraped off upon the scraper, they should be put into a basin, and this is repeated till all the eggs are scraped off and put into the same basin. Water should then be poured upon the eggs, and they should be lightly washed, to separate them from each other. The water will be very dirty, as the eggs are always more or less soiled with the matter deposited by the moth. On the surface of the water will be seen floating the shells of a few eggs, that have already cast their worms (Chap. V.); also many yellow eggs, which are those not impregnated, and others which without being of that colour are very light. All those that float should be skimmed off directly. If the eggs are collected in an unfavourable season, particularly during cold weather, many yellow eggs, and even reddish ones, will sink to the bottom, although they are not impregnated (Chap. IX. and X.) The water having been well stirred, it should be poured into a sieve, or upon some cloths, to drain off the eggs.

The eggs in the sieve, and any that may have remained in the water, are then put into another basin; some sound light wine, either red or white*,

* I have washed the eggs, sometimes with water, and common wine, sometimes with pure wine; hitherto I have not been able to observe any difference in the effects of these

is poured over them, and they must again be washed, and gently rubbed, to ensure their separating from one another.

My custom was slightly to shake the eggs in the wine, and then pouring it off quickly, the light eggs were thus carried of, and separated from the heavy; experience has shewn me that these light eggs are equally as good, and I have found that the difference in their specific weight is very small*. When the wine is poured off, the eggs should be allowed to drip, and then spread upon other linen cloths.

liquids. However, the eggs washed in strong deep-coloured wine, in which they have been left to soak for some hours, hatch later; it appears as if a sort of varnish formed itself round the shell, which may retard a little the necessary evaporation of the humours which give rise to the change of the embryo into the worm.

Those who wash the eggs in muddy wine of a dark colour, thus give to the yellow and dusky eggs a red colour, very similar to the colour of the impregnated egg, and by this artifice, any eggs may deceive, by appearing of a good quality.

* The difference of specific weight in the impregnated eggs of the silk-worm of four castings is not perceptible, indeed I think there is none.

I mention those impregnated, because I have found a manifest difference between these and those not impregnated, of a yellow and reddish cast, although they all possessed a greater specific weight than water.

Thus in an ounce of impregnated eggs there will be found 39. 168
In an ounce of reddish inferior eggs . . . 43. 080
And in an ounce of yellow eggs not impregnated 44. 100

Should the rooms have brick floors, the cloth may be spread on the bricks, and changed every five hours. Bricks dry the eggs by absorbing the moisture more quickly, than any other su b stance.

If the flooring is not of brick, hurdles of wicker-work would be necessary, or basket-work tables. In the course of two days the eggs will generally be dry; they should then be put in plates in layers of 7 or 8 tines, and left until it is needful to hatch them, being careful to preserve them from rats. It is essential to place them in a cool dry spot, in about from 46° to 59° Fahrenheit.

All these operations above-mentioned, until the time when the eggs are laid to dry, employ one hour for thirty ounces of eggs. And this is the distribution of this hour. The cloths in which the eggs are wrapped are to soak six minutes in the pail, the water of which should be quite fresh; the cloths drip five minutes; five-and-twenty mi-nutes may be employed in scraping the eggs off thoroughly, and putting them in the basin; five minutes should be passed in washing them, and separating the light eggs; five minutes allowed to let the water run off through the sieve; four mi-nutes for washing them in wine; five minutes to let that run off and drop off, and five minutes to spread the eggs most carefully upon the linen prepared for drying them.

2. *The necessity of determining by the Thermo-*
meter, the suitable degree of heat for hatching
and rearing the Silk-worm.

To produce, maintain, and regulate the degree of
heat necessary in the space allotted to the hatching
and progress of the silk-worm, we must imitate
the botanist in the management of the hot-house,
and employ the thermometer. By this valuable
instrument, we clearly see that it is of less importance that the silk-worm should live in a temperature equal to the heat of its native climes,
than that it should be preserved from violent
transitions, and thus it should be reared in an
even and progressive temperature.

The thermometer (Fig. 4), simple in itself, as it
cannot be affected by the caprice and will of man, is
a certain method of attaining this important object
of an even temperature; and although it is not the
only requisite instrument in this art, as I shall shew
in Chap. VII., I at present shall mention it alone.

We shall require several well-constructed thermometers.

Thermometers are either made with quicksilver
or spirits of wine. Those made with quicksilver
are always the most desirable, because the expan-
sion and condensation of that metal are more exact than those of the spirits of wine. Besides the
thermometers manufactured with spirits of wine,

low priced as they are, are imperfect; the tube in general is of an unequal interior diameter, which may lead into errors. The expense should not be considered, as good thermometers, made with quicksilver, are of the utmost consequence *.

It sometimes occurs, that the tube of the thermometer is loose, and is misplaced on the scale; this may occasion errors. To remedy this inconvenience, several thermometers should be placed horizontally near each other on a table, upon which should be similarly placed a *correct* thermometer; they should be left thus one hour, to ascertain their precise indications, and then the tubes may be either raised or depressed, and regulated by the correct thermometer, and then fixed with a little melted sealing-wax.

Some persons imagine that they can judge of

---

* Besides the thermometer, there is another valuable instrument for the cultivator of silk-worms, the thermometrographe, invented by M. Cluson Bellani de Monza. This instrument indicates the different extremes which have occurred in the temperature in a particular space of time. With this instrument, the cultivator may know, every morning, always the variations which the temperature has undergone in the hot-house; and may thus watch whether he, whom he has trusted with the management of the stove, has done his duty with exactness. The degrees of heat experienced in the hot-house may thus be equally known in a certain time It is needless to state any further the important use of this instrument. It is a public benefit when enlightened men apply the principles of abstruse science to the useful arts.

the temperature of their hot-houses by their senses
alone, without the aid of the thermometer. Science and practice shew, that few things a,re more
uncertain, and less founded.

The exterior sensations, and the disposition of
the body, are often in opposition to the evidence
of the thermometer; thus the moisture or dryness
of the atmosphere, although the degrees of heat
may be the same on the thermometer, cause us to
experience sensations so different, that we might
feel cooler on a summer's day, the thermometer
being at 82° with a cold north wind, than we
felt the preceding day, when a moist south wind
blew, although the thermometer should have been
many degrees lower. Thermometers are therefore
indispensable *.

* In the book published by Mr. Dandolo, in 1816, may be
found the following passage: "The distance marked from
ice to *boiling water* in the common thermometer is too small ;
the degrees are too close, which sometimes misleads. To avoid
this, I have had thermometers made for the use of the hot-
houses, with a longer scale ; the distance of one degree equal
to that of ten degrees'in the scale of the common thermometers. In this manner I have been enabled to divide each degree into five fractions, which may be easily distinguished
even at some distance. And thus the slightest variations in
the temperature of the hot-house may be perceived at once.
These thermometers have a mark to shew the extreme point
to which spirits of wine rise, which spirit must be coloured.
I must add, that I had them made with spirits of wine, because
those made with mercury were too expensive ; however, when
executed by a clever workman, they are very exact."

M. Bellani de Monza, who made these thermometers, having explained them to me, I communicated his description of

3. *Of the Hot-house or Room in which the Silk-worm should be hatched.*

The first use of the thermometer should be in the hot-house which is destined for the hatching of the eggs.

The eggs of these insects are not hatched by means of animal heat, but by the uniform atmospheric warmth, which should surround them entirely.

As it may be more favourable to our interests that the silk-worm should be developed whenever we find it convenient ; and as this insect must be reared in a season, which, in our climates, has not the requisite warmth, it is therefore indispensable to create an artificial temperature suited to its progress.

A small room or space should be preferred to a large one, as it is thus more easy to regulate the heat, and as it also saves fuel.

The small apartment in which I reared silk-worms, was about twelve feet square, and capable of commodiously hatching, not only ten, twenty, or thirty ounces of the eggs, but even two hundred if required. This small apartment must be particularly dry.

them to M. Lagarde, optician at Paris, who has manufactured some on a similar principle ; they may be had of him, Quai de Gèvres, No. 10, Third Story, Paris. He also makes the thermometrographe, which I mentioned in the preceding note.—*Note of the Translator.*

It should contain all the necessary implements that may be wanted; I may be thought too minute in the following details, but this shall not deter me from giving every explanation I may deem requisite in so important an art.

This small apartment should contain—

1st. A stove of a moderate size, not made of iron, because the heat could not be regulated so accurately, but of thin bricks; it must stand out in the room; it is calculated to raise, by degrees, slowly, and at will, with little wood, the temperature of the room (Fig. 5.)

2d. Several boxes or trays, either made of thick paste-board, if they are not large, but if large, of thin boards (Fig. 6.)

The size of these boxes should vary according to the quantity of eggs which are to be hatched; for an ounce of eggs the space of about eight inches square are required. This may give an idea of the size and number of boxes that may be wanted, and we shall see hereafter how useful it is not to depart from this rule. If there are more than six ounces of eggs to be hatched, the box should be of thin board. Pasteboard boxes should be about half an inch deep. The depth of the wooden trays or boxes must, of course, be proportionate to their size. The boxes should all be distinctly numbered.

3d. Some wicker trays or tables (Fig. 7.)

These wicker trays should be placed horizontally against the wall, supported upon two pieces of wood, fastened into the wall. When there are many of these wicker trays to be disposed of, they should be put one above another, with an interval of about twenty-two inches between them. These trays are for the purpose of holding the boxes in which the eggs are to be hatched. The boxes must be so disposed as to allow of easy inspection, that they may be examined as often as necessary. Care should be taken that the wicker trays be not too close.

4th. A spoon (Fig. 8.) ; it is of a shape convenient for stirring the eggs well.

5th. Several thermometers. The thermometers may be hung in various parts of the stove-room, or still better if laid by the side of the boxes, indicating the precise temperature of every part of the stove-room. For it must be observed, the temperature varies in different parts of the stove-room, particularly between the part next the stove and that nearest the door. This observation may be of use, as it may enable the cultivator either to force or retard the silk-worms by some days ; thus to hatch them as the mulberry leaf becomes fit for their food, which in some soils is earlier than in others.

6th. A few light portable trays, with handles, easy to move (Fig. 9.) ; they are useful for mov-

SILK-WORMS. 47

ing the small boxes, which contain the young
worms, or for moving them when they are more
advanced. They should be long enough to cover
the ends of the wicker trays or hurdles, of about
a foot in breadth.

7th. An air hole or ventilator in the floor of
the room (Fig. 10.), with a sliding panel to
open and close it, but which, in general, must be
closed; it may be used to temper the heat, should
it have exceeded the degrees which I shall point
out, as necessary for bringing forth the silk-worm.
We have thus the means of making a gentle current of air between this air-hole and the door,
to correct the excess of heat indicated by the
thermometer.

8th. A glazed window, to light the stove-room;
it is a vulgar error to imagine that light is not as
necessary to the animation of the silk-worm as
to that of every other living thing. The light
does not incommode the silk-worm, until it has
reached its perfect state of moth, as we shall
mention in the tenth chapter.

This is all that is required to furnish a stove-
room. It is superfluous to add that this room
may be used for rearing the silk-worms, as well
as for hatching them; and also that this room,
which may be heated with little expense, might
hatch silk-worms for any number of persons: it
might be considered as a species of shop, which

receives, works, and sells, in the usual course of trade *.

* Without quoting other years in which the spring was warmer than in 1814, I will give an account of that year, in which I was obliged to retard, by some days, the hatching of the silk-worms. I will note the temperature of the stove-room, and the exterior temperature of this western room, at five o'clock, A.M.

Temperature of the Stove-Room.

May.	Degrees. (Fahrenheit).	Exterior Temperature.
11	64	53
12	64	46
13	64	46
14	64	46
15	66	48
16	66	53
17	68	51
18	71	51
19	73	51
20	75	53
21	77	53
22	80	55
23	82	53

During these thirteen days, employed to prepare and obtain the hatching of the silk-worm, we used two quintals, one pound, of wood, large and small. I have already stated, that the brick stove is more convenient than that of iron, which consumes a vast quantity of fuel very speedily, and the heat of which, as it cannot be well regulated, may kill the silk-worms. I might have forced these eggs some days earlier, but I was obliged to keep them back, as the unfavourable weather had retarded the springing of the mulberry leaves. However, even thus, I gained three or four days upon the time ordinarily fixed for the hatching of the eggs in Chap. IV. § 4.

4th. *Of the Birth of the Silk-worm.*

When the cultivator has observed the state of the
vegetation in the mulberry leaf, and imagines it
fitting to have his silk-worms hatched in ten days,
he will put the eggs in the boxes in the proper
quantities; he must weigh them carefully, and
keep a register, in which he must note his observations upon the course and progress of the insects, and thus at once secure theory and practice; beginning by marking the day and hour
upon which he sets the box in the stove-room,
and also the number of the box, and, in short,
every thing that may be worthy of notice. The
wicker hurdles should have paper laid inside of
them, and the distance I have desired may be left
between the boxes, is to prevent the silk-worms
from going from one to the other.

If the temperature of the stove-room should
not reach 64° on the day fixed upon to put in the
eggs, it is necessary to light a little fire, that it
may rise to that degree of temperature (64°),
which ought to be continued during two days.
Should the thermometer indicate that the exterior air is above 64°, the shutters should be
closed, and the door and the ventilator opened to
create a draught, and cool the stove-room. The
third day the temperature should be raised to
66°; the fourth day to 68°; the fifth day to 71°,

D

the sixth day to 73°; the seventh day to 75°;
the eighth day to 77°; the ninth day to 80°;
the tenth, eleventh, and twelfth days to 82°.

The following are the signs of the speedy vivification of the silk-worm.

The ash-grey colour of the eggs grows bluish,
then purplish, it then again grows grey, with
a cast of yellow, and finally of a dingy white.
These shades of colour will vary, and they depend also on the means used in washing the eggs;
I have seen them so deeply stained by red wine,
that the colour remained unaltered even after
the worm had cast the shell.

If the eggs of silk-worms, belonging to different proprietors, are put into the same stove-
room, great differences will be observable, not
only in changes of colour in the eggs, but also
as to the period of hatching the worms. The
insects of the eggs, that have been preserved
through the winter, in an even and gentle temperature, and those of the eggs which have undergone maceration*, come forth four or five

* By maceration is commonly understood, eggs preserved
in bags, under cushions or mattresses, or in blankets and
similar things, until the moment of putting them in the
stove-house. Those who thus macerate them, are careful
to move them now and then, to prevent their overheating;
maceration is employed to ensure their speedy hatching when
put into the stove room, or elsewhere.

After this, can the cultivator have any idea or certainty,

days sooner, namely, at the 71st, 73d, or 75th
degree of temperature; whilst those that have
been kept in a very cold atmosphere appear some
days later.

This stove imparts to each egg the degree of
heat necessary to change the embryo it contains
into the worm. When the eggs have been kept in
a certain degree of warmth, it requires less stove
heat to develope the silk-worm. This is so true,
and so worthy of notice, that we find, if in the
winter the eggs have been kept in an atmosphere of 55° or 59°, or heaped together, they
come forth without the aid of the stove spontaneously, when the room is but slightly warmed,
and before the mulberry-tree has given any sign
of vegetation. In this case these worms must be

as to the degree of temperature to which the eggs have been
exposed, and how many degrees may be wanting to complete
their vivification in the stove-room ? Can it be calculated
how high the degrees of the thermometer should be raised in
the stove-room, to receive these macerated eggs, without
injuring the embryo, or the progress of the silk-worm?
This uncertain method must needs be injurious to the regular
and secure development of the worms. I have myself often
seen great quantities spoiled by maceration, the worms
coming forth, and shortly dying.

It appears to me rational that, when there appears to be
a certain and regular method, we should not adopt another,
the result of which is uncertain and irregular, particularly
when this secure method is not attended by any exorbitancy
of expense, as I have just shewn, in the preceding note, and
in Chap, IV. § 4.

thrown away. This circumstance is, therefore,
of essential consequence, and should be noted, to
prevent its occurrence. A little delay in the
hatching of the silk-worms is no loss; whereas
it is a very serious loss, if they anticipate by a few
days the proper period of hatching. To backen
them when very near the time of coming forth,
by altering the temperature, injures them materially. (Chap. XII.)

When the egg assumes a whitish colour, the
worm is already formed, and with a glass may be
seen within the shell. The eggs should then be
covered with white paper, well pierced with a
particular instrument made on purpose, (Fig. 11.),
the paper cut so as to cover them all. The
worms will appear upon this paper climbing
through the holes : a clear muslin will do as well
as paper. To collect the worms, small twigs of
mulberry, with only two or three leaves on them,
should be laid on the paper, and they should be
increased as fast as the worms come out upon
them, for if they do not find the leaves they get
out of the boxes. Few worms appear the first
day, and if the number of them is very inconsiderable they should be thrown away, because
when mixed with the later worms they would
grow faster, and becoming mature so much sooner,
they would only be troublesome. I prefer the
small spray of the mulberry to the flat leaf, as I

have observed the leaf is apt to press upon the young worm, and load it. Many cultivators may have witnessed, that when I removed the old leaves there 'were a number of worms injured, not being able to disencumber themselves from the weight.

The worms which may have been managed according to the method I have stated, will always be healthy and strong. They will neither be red nor black, but of a dark hazel colour or chestnut, which is the proper colour they should have.

It is impossible to state the practical advantage of this method, thus ensuring the constant production of well-constituted animals *.

* The essential point is to cause the eggs to be hatched with the greatest ease : if the success of this operation is not complete, the worms will probably be subject to disease through their whole course of life, as I show in Chap. XII.

It has been seen, in the two latter notes, how necessary it is to make use of the brick stove, and that the expense of heating the room for a certain number of days is inconsiderable. Would it not, therefore, be most useful to establish, in those countries where silk-worms are cultivated, public stove-rooms, and also rooms to put the worms in when hatched, previous to distributing them among the proprietors and farmers? It would be a more secure, more certain, and much less troublesome method for those who are in the habit of rearing silk-worms in the old way. For one hundred francs, you might thus hatch thousands of ounces, and thus begin to nationalize (if I may use the expression) this art, which is the main spring of numberless other arts.

In cases where the public would not incur the expense, might

The appearance of the new-hatched worms is
that of a woolly substance, of a dark chestnut
hue, in which is perceptible a general stirring of
minute animals, rearing up their heads, and presenting a black and shining speck, or head. Their
bodies are covered with regular lines of hair, or
down, of various length. Their brown colour is
caused by this hair, the skin being whitish, which
shews as they grow larger and the hair thinner.
The whiteness of the skin is perceptible the mo-

it not be defrayed by the proprietor of seed paying a certain sum for the use of the stove-house, proportionate to the
expense incurred?

The utility of this establishment would be considerable,
particularly if the person conducting it was well instructed
in the art of managing the silk-worm, and would communicate his knowledge to the mere drudging labourer. It
would diminish the losses caused by ignorance and error.
If the world is overrun with quackery and erroneous systems, why should not good, enlightened, and patriotic
men strive to benefit mankind, and enable them by true
science to cultivate arts calculated to increase their riches
and happiness? It is with the hope of seeing such men
spring from various points, to raise and protect such institutions, that I express myself thus warmly, and they will
have their reward in the blessings of generations to
come [a].

[a] What the author has stated has given me the idea that the
States-General of the Departments might propose to government these institutions as the means of increasing public industry, and thus to adopt in each commune an establishment of
this description, for the hatching of the silk-worm in common.
It is not to be doubted, that a vast saving of the cocoon would
thus be ensured.—(*The Translator.*)

ment it casts its shell and stretches its head. When seen through a glass, it appears to have a white collar ; the tail is also bristled with hair.

While the eggs are in the stove-room, they should be stirred round with the spoon two or three times a day; this operation is useful, and, besides, hastens their coming forth.

When the temperature of the stove-room is raised to 75°, it is advantageous to have two dishes, in which water may be poured, so as to offer a surface of nearly four inches diameter. In four days there will have taken place an evaporation of nearly twelve ounces of water ; the vapour, which rises very slowly, moderates the dryness which might occur in the stove-house, particularly during a northerly wind : very dry air is not favourable to the developement of silk-worms. (Chap. XII.)

When carefully following the precepts I have given, I repeat it, there will invariably result the certainty of obtaining silk-worms, healthy and strongly constituted.

Thus the incubation of the silk-worm being concluded, let us advert to the preparation of the habitation it is soon to occupy, and of the means of removing them ; we shall then return to the subject of this chapter.

CHAPTER V.

OF THE SMALL LABORATORY TO WHICH THE NEW HATCHED
WORMS SHOULD BE REMOVED; OF THEIR REMOVAL, AND
ON THE PROPORTION BETWEEN THE WEIGHT OF EGGS
AND THAT OF THE WORMS OBTAINED.

EXPERIENCE constantly shews, as we shall
prove in Chap. XII., that if it is injurious to expose the silk-worms to a great heat and dryness
of temperature, it is no less so to transport them
into a cold atmosphere, were it only for the space
of one or two days. (Chap. XII.) And facts
equally demonstrate the usefulness of proportion-
ing the space of the rooms to the quantity of
silk-worms it is intended to rear, both for economical motives, and also for the well-being of the
silk-worm: it is, therefore, necessary to determine the space which a certain number of silk-
worms should occupy, during the several stages
of their progress. It is also worth observation to
remark the progress by which in the evaporation
of the egg, the embryo becomes a healthy and
strong silk-worm.

We shall, therefore, in this chapter treat of the
following subjects:

1. Of the laboratory, or closet destined to receive the new hatched silk-worm.
2. Of moving the silk-worm, directly after it
   is hatched.

3. Of the loss of weight experienced by the
eggs in the stove-room.

1. *Of the Laboratory, destined to receive the Silk-
Worms newly hatched.*

This small habitation is to contain the silk-worm
until its third casting of skin, or moulting *. The
room should be in exact proportion to the number of silk-worms, and calculated for facilitating
the attendance upon them. Thus proportioned it
will be economical, as there will not be that
quantity of fuel used, which it would take to
warm either one very large room, or several
small ones.

It may be necessary to know the space required
by silk-worms, in their various stages of growth.
The worms proceeding from an ounce of eggs re-

* In explaining the use of the small laboratory, I have in
view to show how much more economical it is, than apartments either too large, or indeed too small. However,
others must be guided by their own convenience in making
use of those rooms they find suitable. And should there be
but one apartment for the rearing of the silk until the cocoon
was formed, it would be of small consequence, provided attention were given to maintain with exactness the temperature, in all its degrees, which I shall indicate. One room
suffices, particularly for those who hatch but a small quantity of silk-worms, if they have wicker tables enough to put
sufficient space between the worms. These should be two
hundred and ten feet square, for disposing of the silk-worms
proceeding from one ounce weight.

D 5

quire, until the first moulting of the skin, a space of about seven feet ten inches square.

Until the second moulting, a space of about fifteen feet four inches square.

Until the third moulting, a space of thirty-five feet square.

The hurdles, or wicker trays, should be placed above one another, at a distance of twenty-two inches at least; and as many should be put as are required to give the number of square feet necessary for the quantity of silk-worms which are to be accommodated.

The worms must be kept upon paper, which should line the wicker trays, and extend a little beyond, to prevent the worms falling off.

Upon this paper, which should be rather strong, ought to be inscribed the corresponding numbers to those on the boxes; thus avoiding every chance of the silk-worms of different boxes being mixed.

According to the size of the room, there should be one or more thermometers. One or two small fire-places placed in the angles, one or two ventilators in the ceiling, or in the floor, and one or more windows; and also as many doors as may be convenient. In this room I also would recommend a stove similar to that in the stove-room, as in cold weather it may be of use to save fuel.

It requires more wood to heat a room during one day, even with the best-constructed fire-places and chimneys, than to heat the same room for ten days by means of a stove. The principal advantage of the small fire-places is not so much its warming the air, as its making a draft or current through it, as we shall shew hereafter. (Chap. VI. and VII.)

The temperature of this laboratory should be carried to 75°; about 2° lower than the stove-room heat which hatched the eggs.

Experience teaches us, that as the silk-worm grows older, and gets stronger, it requires less heat.

Such is the temperature that suits these insects shortly after they are hatched. Should the season be peculiarly unfavourable, and the vegetation of the mulberry-tree checked, it might be necessary to slacken the temperature, and thus gain a few days by gradually lowering the heat to seventy-one degrees, and even to sixty-eight degrees, but not beyond that *.

* A prudent proprietor has done all in his power, when, on observing the season favourable, and the bud of the mulberry shoots in a proper degree of forwardness, he has put the eggs into the stove-room. Should the weather suddenly change, as it did in 1814, it is then of great use, to have the power of backening the hatching of the eggs without injuring the worm, as I have before stated, and to prolong

2. *Of the removal of the new-hatched Silk-Worm into the Laboratory, or elsewhere.*

The silk-worms should be removed as soon as possible into the laboratory, in which they are to

their two first stages by a few days. To obtain this, the
only method is, after the worms have been removed into the
laboratory about five hours, to lower the temperature to
73° from 75° ; four hours after, further to lower it to 71°,
and the following day to 68°, if necessary.

This cooling of the air diminishes the hunger of the young
silk-worm by degrees, and without danger; and by these
means the modifications are prevented, which, at 75°, would
have brought on the casting, or moulting, much more
speedily.

At 75°, the first moulting is effected the fifth day ; whilst
at 71°, it requires six or seven days. The second moulting, which at 75° is wrought in four days ; at 69° and
71°, takes six days for its accomplishment. Thus, by
foresight and prudence, the proprietor will be enabled to
gain seven or eight days, which prevents any ill effect from
the unfavourableness of the season ; and this time gained, it
is evident, may be of the utmost consequence. The tables
I have annexed, at the end of this volume, will shew that, in
1813, the silk-worms were reared in thirty-one days, and that
it required thirty-eight days to raise them in 1814, to allow
time for the growth of the mulberry-leaf; and I do not
comprise in these seven days which I gained, three days
which I delayed in the hatching of the silk-worms, having
perceived that the whole season was bad. Those who are
not careful thus to meet the accidental untowardness of seasons, and by art to prevent their injuries, would be obliged
either to throw away the early-hatched eggs, or to strip the
mulberry-tree too soon, and injure the leaves which are to feed

remain until their third moulting unless it is intended they should be reared in the stove-room. If I propose a different room for the rearing of these insects, until their third moulting, it is because I found it both more convenient, and more advantageous.

When they are about to be removed from the stove-room, three circumstances must be noted, relative to the manner of transporting them.

In the first place, whether the silk-worms are to be reared in the same house in which they have been hatched.

Secondly, if part of them are to be reared in; the house, and the rest removed.

And lastly, if all are to be removed.

1. Let us suppose all the worms are to be reared in the same place.

When the little twigs, spread over the perfo-

the silk-worm in its adult stages hereafter. These considerations must strongly impress the necessity of delaying the hatching of the eggs by some days, rather than hurry their coming forth ; particularly as there is no fear, when worms are reared in this secure manner, of their being injured ; should there occur two or three hot days, these would only accelerate the moulting a few days sooner. It is also certain that the later silk-worms, in their last stage of progress, make choice of the leaves suitable to their age, and particularly those leaves which are quite ripened, which, for the proprietor's interest, is the most important period, as it is at that last period the greatest consumption of the leaves occurs.

rated paper which covered the eggs of the silk-
worms in the small boxes, are loaded with the
young worms, these boxes are all put upon trays
made to carry them, and they are removed to the
laboratory.

When all the sheets of coarse paper, accurately
numbered, have been laid on the wicker hurdles,
the small box of the corresponding number is put
on the edge of the wicker hurdle, and with a
small hook, the twigs, covered with the worms,
are gently lifted off the perforated paper on the
box, and put upon the paper laid on the wicker
hurdles. The hook should be used, (Fig. 12.) as
the touch of the hands might injure the insects. In laying the twigs on the paper, care
must be had to allow space enough for mulberry-
leaves to be put over the twigs, and between
them, that the insects may have room to stretch
and properly distribute themselves. It should be
noted here, that the silk-worms produced from one
ounce of eggs thus disposed, should occupy a
space of about twenty square inches. Each sheet
of coarse paper, on the hurdle, will cover a space
of nearly twenty-two square inches; being twenty-
three inches long, and twenty-one inches wide.
Having a care to lay the worms in small squares
of ten inches, four sheets of paper will be the
number required to hold the produce of one ounce
of eggs, which will exactly allow the worms the

space they need until after their first moulting. The sheets of paper will be four times the size of the small boxes, and those four sheets of paper must bear the same number as the box ; and thus the worms will not want moving till their first moulting is passed.

As fast as the silk-worms come forth, they should be moved in this manner *.

* It is easy to imagine that it may often require more than three days even to bring forth the silk-worms from a given quantity of eggs. It will be seen in Chap. X. that the moths do not issue from the cocoons in less than ten days or a fortnight, according to the temperature to which they have been exposed ; and it is therefore evident there may be a difference of ten days or a fortnight in the laying the eggs. As the eggs put to hatch are therefore not all laid the same day, and are liable to the same degree of heat in the stove-room, some must come out sooner than others[a] ; hence no one can say the late-hatched eggs can be either better or worse than the early eggs, because the embryo has required longer to perfect itself into the worm ; this period is always proportioned to the constitution of the eggs. These reflections should satisfy the small proprietor, who has but one box of

[a] It does not appear this should be exact evidence, that because the eggs are not all laid in one day, they cannot be hatched in one day. If we may argue by analogy, it is well known that hens hatch eggs laid at various periods in a short time. Housewives well know, when they choose eggs for setting, that, provided they be good eggs, their having been laid at different times is of no consequence. It would appear, that it is not because the eggs of the silk-worms have been laid sooner, or later, that they do not hatch at one time ; but, more probably, that this difference proceeds from the peculiar quality of the egg, and of the care taken to surround it constantly with the degree of heat it may individually require.

When they are laid upon the papers, they
should be given some young leaves chopped

eggs, and one single room to rear the worms in, how much
it imports that he should not reckon on the very late-hatched
eggs, not from any defect in the quality of them, but that he
may not have worms of a day old mixed with worms four
days old, thus interfering with the general progress of cultivation.

The proprietor, on the contrary, who has many boxes of
eggs to bring forth, can dispose of the early eggs to his tenants, and by these means need never mix the silk-worms of
different ages. Then, if one tenant holds those of the first
day's hatching, and another holds the worms hatched the
fourth day, no evil accrues, all proceeds with regularity and
ease, as each tenant has equal-aged silk-worms to attend to.

When the proprietor has only a small box of eggs to hatch,
it is better to cast away those that come forth the first day, and
not to reckon on those that are not come forth the third day;
and thus, by having only those hatched in the two intermediate days to attend to, much trouble will be saved.

If in this case the cultivator wished to act with that theoretical exactness, which is the ground of every art, and if he
wished to know what really is the quantity of worms he
rears, when the third day's hatching is over, he should weigh
the eggs that remain unhatched, adding one-twelfth to their
weight which they had lost in the stove-room, as may be seen
in Chap. V. § 3.; and thus he will exactly know the
effective quantity of eggs to which the worms correspond.
In general, the worms hatched the first day are very
few; but calculating upon this view, that sixty-eight silk-
worms are equivalent to the weight of an egg, the proprietor, upon this calculation, might throw them away if there
were not many hatched. It is far better to lose a few worms
of the first day's hatching, and eggs not hatched the third
day, than to be inconvenienced during the whole progress
of rearing them; by adding a small quantity of eggs to
those that are next to be hatched, the loss is easily made up.

small, covering the spaces between the twigs with
these chopped leaves, that by degrees the whole
surface may be equally spread with silk-worms.
In case they should get into heaps, a leaf might
be put over them, to which they will adhere;
and, being gently lifted, it may be put in any
spot where the worms lie thinner.

Whenever silk-worms are added upon the paper where some worms have already been deposited, they should have food given them; but the
worms that were first laid on the paper should
not be again fed until the other sheets of paper
have been filled. Thus a fair number of the
first worms will receive the second meal at one
time.

The worms take at least two days to come forth,
consequently the first hatched will be larger than
those that appear the second and third day. We
have stated above, the thermometer proves, that
a heated room can never be heated in a degree
exactly equal in every point of space. There will
exist the difference of a degree, and even more :
we have also said, that the vicinity of the door,

I recommend these directions to be exactly followed up;
they will guide, simplify, and ameliorate the cultivation of
silk-worms. If they are not attended to, it will not be
known which is the exact proportion of eggs which produces
the worms, and there will be constantly on the tables worms
of different sizes with differing necessities, and failure and
trouble will ensue.

and that of the stove, constitute the two extremes. By putting the early worms in the coolest part of the room, and the late-hatched worms in the hottest, and by feeding the latter rather higher, it is practicable to bring them nearly to an equality.

2. We are to suppose that a portion of the silk-worms are to be reared in the house in which they have been hatched, and the rest to be reared elsewhere.

I have said all that is necessary respecting the silk-worms which are to be reared at home. I shall only mention, as to those that are to be removed from the laboratory, that, to facilitate their conveyance, each sheet of paper should contain the whole ounce of silk-worms, and not the quarter alone. On each sheet should be disposed, in a space of eighteen inches square, which when filled will hold about the whole ounce.

When the cultivator shall have taken home with him his portion of silk-worms upon these sheets of paper, containing one ounce, he must put the square of silk-worms into four small squares of ten inches each, thus forming four quarters of one sheet, or rather four small sheets. This division is easily effected, by passing the hands under the litter of leaves to which the worms cling, and parting the leaves gently ; when it is required to divide the mass, it separates easily, and may be subdivided at pleasure. The parts should be as even as possible.

If, in the first ages of the silk-worms, all the care I have described is not taken, numbers will be lost (Chap. IV.); they will be unequal, and contract numberless diseases.

3. We are to suppose all the silk-worms are to be removed from the house in which they were hatched.

The rules prescribed for the removal of part of the silk-worms is applicable to the removal of the whole of them; but, as it may be necessary to remove them to considerable distances, it requires some particular care, which is however very simple.

In a box, calculated for removing easily, (Fig. 13,) and proportioned to the size of the sheets of paper, should be put several of these sheets, covered with silk-worms in layers above each other, at the distance of nearly two inches. This box may be carried on the back, like a pedlar's pack, with little fatigue, although it should contain several ounces of silk-worms. Should this box, which however I think is useful, not be employed, common baskets may be used*, which would not

* The *hotte*, or high basket, which is here understood, is used in France and Italy by the peasantry, and carried upon the back by means of straps, through which the arms are put; but I do not see why our common baskets should not answer as well.—*(Note of the Translator.)*

carry more than four or five ounces at a time. The removal in baskets may be executed in safety, with the following precautions :—

1. Lining the basket thoroughly with paper well-closed, that the exterior air may not strike the silk-worms, particularly if it should be cold.

2. Preventing the sheets of paper covered with worms from touching each other, by putting slender sticks across to support the sheets of paper, and avoiding their pressing together. This should be done in as many layers, from the bottom of the basket to the top, as there are sheets of paper covered with the young worms, leaving a distance of four fingers between each.

3. To cover the basket very completely with linen cloths, to keep off cold and sun.

4. To remove them between the hours of twelve and three o'clock, that being the hottest part of the day.

5. To give the worms a small quantity of young chopped leaves, if their journey is likely to be three or four hours long.

I trust I have clearly, and with simplicity, shewn what it is requisite to do for the hatching of the silk-worms, and the removal of them into the laboratory, or elsewhere ; and the active cultivator will, I hope, find the execution of my directions easy and plain, provided he has duly disposed all the things he may require. When once the laboratory is established it lasts for life.

The proportion of the boxes which I have stated as necessary for hatching the silk-worms should not be altered; as this proportion prevents any necessity of ever touching the eggs from the moment the silk-worms begin to appear.

The perforated paper, being large enough to support a number of the small twigs of the mulberry, it consequently enables us to remove a large portion of the silk-worms at once. In using these small boxes, the egg-shells will always adhere together, and when the boxes are lifted they should be slightly shaken horizontally, to move the eggs. If in moving the eggs some of the holes in the paper should be stopped up with the eggs, it is of no consequence, as it will not prevent the worms climbing up. Should there be any wish to see the species of tissue which unites the silk-worms to the eggs, by gently lifting the perforated paper it may be examined, but the paper must be carefully replaced. Before the appearance of the worm, even the eggs adhere, and are fastened together by a substance, or exudation of the egg, which cannot be discerned even with glasses.

Whenever a sheet of paper is prepared for the arrangement of silk-worms, there should be inscribed upon the paper itself the hour in which the arrangement began ; thus it will be seen in what time, and in what progression, the silk-worms come forth. A pencil may be worn, for the purpose of noting the sheets of paper.

I have said, that if the worms which appear the first day are in very small quantity, as it mostly happens, it is of no consequence, because the main portion come forth the second or third day. However, if it is required that those first hatched should be reared, they ought to be put in an angle of the sheet of paper belonging to the number of their box, and only be allowed half the quantity of the food which is given to those later worms on the first and second day.

It appears, in general, that the silk-worms come forth more abundantly in the forenoon, when the sun shines warmly into the room, the room being then hotter than at night. Those who have the care of the silk-worms at night drop asleep ; and when, in those cases, I have gone into the stove-room, I have often found the thermometer lowered by some degrees ; it is how-ever better it should lower a degree or two, than that it should, by neglect, rise too much. Sudden change of temperature injures the embryo which is near hatching. (Chap. XII.)

The great alterations to which the eggs are ex-posed occur in the night. Those who have the care of the silk-worms at night heap up the fire,

that they may take their rest without having to
make it up, which augmentation of heat affects,
and even spoils, the embryo.

I have observed that, some days, the hatching
of the worms was most abundant in some boxes,
and equally so in all the hours of the day as in
the morning.

It may not be out of place here to make known
an easy and beneficial improvement which might
be made upon our local customs.

We have proprietors who hatch large numbers
of silk-worms, afterwards distributing to their
tenants, in small boxes, in proportion to the quantity of food for silk-worms which the tenant has at
his disposal.   Instead of which I would propose,
that all the eggs should be hatched in boxes, capable of holding twenty or thirty ounces, constructed in the proportions before-mentioned, and that
as fast as the worms come forth, the sheets of paper should be arranged to receive the ounce of
silk-worms in regular order, as I have also before
explained ; by this method, each tenant would receive worms hatched nearly the same hour, perfectly equal, and that may be easily reared, as
experience has shewn.   When all the silk-worms
are hatched, they should be divided into ounces,
as nearly as possible, and put upon the sheets of
paper.   The earliest should be given to those
cultivators whose mulberry-trees are most ad-

vanced. Should the hatching continue three days upon this plan, it would make no difficulty, as different tenants would take the worms of various ages, and thus each would have the silk-worms of one period. It is a great error to imagine that it can be advantageous to give a tenant silk-worms of various days hatching to make up the quantity he is to receive and rear, because those of the first day's production are stronger than those of the second day. We repeat it, the essential point is, to have the worms as nearly equal as we can bring them.

3. *Of the Diminution in the Weight of the Egg, before the hatching of the Silk-worm.*

I ought perhaps here to speak of the treatment of the silk-worms when removed into the laboratory, but it may be better to add some further observations which I made; although not directly relative to the art of rearing the silk-worms, I trust they may not appear quite irrelevant.

The loss in weight of various thoroughly dry eggs, when placed in the stove room, is as follows, beginning by a temperature of 64°, as I before mentioned. (Chap. IV. § 4.)

8 ounces of eggs lost *grs.* in weight in 5 days	*grs.*		*grs.*	
in weight in 5 days .	100	In 8 days . 360	In 10 days .	440
6 oz. lost in 5 days .	86	In 8 days . 178	In 10 days .	248
5 oz. lost in 5 days .	60	In 8 days . 168	In 10 days .	216
4 oz. lost in 5 days .	80	In 8 days . 181	In 10 days .	224
In 5 days .	326	In 8 days 887	In 10 days	1128

Boxes containing the same quantity of eggs, and
even smaller ones, lost nearly as much in weight.
In five days, the evaporation of the eggs in the
stove-room, is 13 grains per ounce; in eight days
37 grains; and in ten days, which is till the period
of hatching, 47 grains. Thus one-twelfth of the
weight of the egg evaporates previous to hatching. The shells or husks of 24 ounces of silk-
worms gave the following weight:

A small box of eight ounces, gave 1020 grains of husks.
Ditto   six ounces, gave   724   ditto   ditto.
Ditto   five ounces, gave   504   ditto   ditto.
Ditto   five ounces, gave   548   ditto   ditto.
                                 ————
                                 2796

The average weight of the shells or husks is thus
equivalent to about 1-5th of the weight of the eggs.

To make one ounce of picked eggs, there should
be for an average weight 39,168 eggs. I observed, with some surprise, that there was little
difference in the weight of eggs belonging to
above twenty different persons.

I have had the patience to count many hundred
thousand eggs, in hopes that it might be useful in
the art of rearing silk-worms. The best eggs,
when weighed, afforded no more than 68 eggs per
grain. And the inferior quality of eggs did not
afford more than 70 eggs per grain.

I will only add here, that about 360 good cocoons weigh one pound and a half. That those
who suffer no loss, either in the eggs, or in young

E

worms, might obtain from one ounce of eggs a hundred and sixty-five pounds of cocoons ; and whatever less is derived from this quantity may plainly shew the exact amount of loss and damage.

One ounce of eggs, composed of 576 grains, is reduced to 413 grains, deducting 47 grains, lost by evaporation, and 116 grains in the weight of husks. The 413 grains are thus equal in weight to 39,168 young worms. At this rate 54,526 young worms, newly hatched, are required to form the weight of one ounce.

Attentively examining the different facts relative to the varieties in the quality of eggs, I have convinced myself, that particularly cold weather, such as we had in 1813, at the time of the birth of the moth, much injures the impregnation of the eggs. Among all the different qualities of eggs I then examined, I found not one that did not contain in weight a proportion of $\frac{1}{12}$ and even $\frac{1}{8}$ of yellow and dingy, or unimpregnated eggs. I carefully chose 5000 yellow eggs, and 5000 red eggs; they had all a specific weight greater than water, as they sank when washed. I had them put into a box, and placed in the stove-room, with the other boxes of eggs; one single worm alone was hatched from a red egg. They remained filled with matter ; but as they were not impregnated, they could produce no worms. They diminished more in weight than the impregnated eggs.

Second Table.—REARING of SILK-WORMS proceeding from Five Ounces of Eggs.

1814 Days of rearing	Months	Sorted leaves	Internal temperature bells	External temperature devoted to the rearing, vertical	Hygrometer M. Rolland	Weather	OBSERVATIONS

FIRST AGE.

The silk-worms, on some tables or hurdles, prompt sooner than others; by reason at the exterior cold, the temperature of the small Laboratory was, in some years, a degree and a half below other parts, although the waste Laboratory was well closed, and fitted to keep out the cold. This degree of cold was on the side of the spectrum, and hotter ones at tray hurdles.

SECOND AGE.

The silk-worms became turgid and roused with most regularity, and at, less distant, periods than in the first age.

THIRD AGE.

All proceeded with regularity in this third age. There came twenty-four pounds more of mulberry leaves consumed this year than in 1813. There eat ten refuse picked from the leaves in 1814 than in 1813, consequently, the average quantity of leaf meal have been nearly the same both years.

FOURTH AGE.

Two days were tapping out in counting the wickers, because the silk-worms that were placed in the coolest part of the Laboratory, became torpid, and roused a whole day later than the others.

Talley, thirteen more of sorted leaves were consumed than in the year 1813, but in 1814 there was less refuse. The progress of the fourth age was tolerably regular.

FIFTH AGE.

The easy and wastefulness of silk-worms, render these last eleven days of the lesson remarkable.

The silk-worms continued to prosper. But as the eighth were very cold, an oven temperature throughout every part of the Laboratory could never be obtained. Fires were lighted in the stoves, and each wood was burnt in the grates, to maintain the necessary temperature.

There were eighty-four pounds more of leaves consumed than in the year 1813.

The refuse of the leaves and the weight of dung were less than in 1813.

There were fewer mulberries on the branches, and even those were kinder in 1814, than in the preceding seasons. Those worm obtained six pounds more of cocoons. Some worker families composed a little less than the 28th or 29th day, and their produce was inferior.

1814 Pounds of leaves per ounce of eggs.——The silk-worms of five ounces of eggs having consumed 2139 pounds of leaves, produced 604 pounds 6 ounces of cocoons approximate, and 3 pounds 8 ounces of offals. About 20 pounds of leaves were considered for a pound of cocoons.

The material originally positioned here is too large for reproduction in this reissue. A PDF can be downloaded from the web address given on page iv of this book, by clicking on 'Resources Available'.

FIRST TABLE.—REARING of SILK-WORMS proceeding from Five Ounces of Eggs.

1813 Days of rearing	Months	Sorted Leaves given to the Silk-Worms	Internal temperature	External temperature	OBSERVATIONS.
1st Age Day 1	May 18	*lbs. oz.* 4 4	75	0	**FIRST AGE.**
2	19	6 0	75	64	The day of rearing begins after the twelve o'clock afternoon.
3	20	12 0	75	65	Dry cold air.
4	21	6 4	75	66	The exterior temperature was at the 22nd to 6th degree, to the west at five o'clock in the morning.
5	22	1 8	75	65½	
Leaves		30 0			
2nd Age Day 6	23	18 0	73½	64	**SECOND AGE.**
7	24	30 0	73½	64	Fine weather.
8	25	33 0	73	64½	Dry air.
9	26	9 0	73	66	
Small branches and leaves		90 0			
3rd Age Day 10	27	30 0	71½		**THIRD AGE.**
11	28	90 0	71½		Little variation in the external temperature, south winds predominated, heavy air, rain.
12	29	97 8	71		The dry air experienced in the second age had imparted great vigour to the insects.
13	30	52 8	71		
14	31	30 0	71		
Small branches and leaves		300 0			
4th Age Day 15	June 1	0 0	71		**FOURTH AGE.**
16	2	97 8	71		
17	3	165 0	69		
18	4	225 0	69		Some silk-worms placed in a temperature of 64 degrees became torpid during fifty hours.
19	5	255 0	69		
20	6	127 8	69		
21	7	30 0	69		
22	8	0 0	69		
Small branches and leaves		900 0			
5th Age Day 23	9	180 0	69		**FIFTH AGE.**
24	10	270 0	69		The season was extremely unfavourable from the 13th until the 18th of June. The thermometer placed in the
25	11	420 0	69		west, stood during the 15th, 16th, and 17th as low as 33 degrees, at five o'clock in the morning. Cold rain almost
26	12	540 0	68		uninterrupted, although the barometer did not indicate great moisture of atmosphere.
27	13	810 0	68		
28	14	975 0	68		
29	15	900 0	69		
30	16	660 0	69		
31	17	495 0	69		
32	18	240 0	69		

Weight of mulberry leaves per ounce of eggs 1609 pounds 8 ounces. The silk-worms proceeding from five ounces of eggs, which have consequently consumed 8047 pounds 8 ounces of leaf, produced 667 pounds 8 ounces of choice cocoons, and six pounds of coarse floss. About 90 pounds 1 ounce of leaves were consumed to a pound of cocoons.

The material originally positioned here is too large for reproduction in this reissue. A PDF can be downloaded from the web address given on page iv of this book, by clicking on 'Resources Available'.

Of good eggs, there does not remain at most one hundredth part that is not hatched in the course of the three first days. This hundredth part continues to hatch in time, but should not be calculated upon.

These observations may be useful to those who like to know every branch of the art of rearing silk-worms; they are at least curious, and as far as I know, they have the merit of appearing new.

CHAPTER VI.

OF THE REARING OF SILK-WORMS IN THEIR FOUR FIRST STAGES.

LET us now touch upon the subjects more particularly relating to the silk-worm.

In the preceding Chapter, it has been stated, that the space suitable to the number of silk-worms proceeding from one ounce of eggs, should be, in the first age, that is to say, until the first moulting, 7 feet, 4 inches, square.

Of about 14 feet, 8 inches, until the second moulting; and of 34 feet, 10 inches, till the third moulting. The space required until the fourth moulting, should be of 82 feet, 6 inches, square.

Those who have the means of extending these allotted spaces, may do so by some feet, because it is certain that the more room silk-worms are allowed, the better they eat, digest, breathe, perspire, and rest. The spaces I have mentioned are sufficient, and present the advantage of facilitating the attendance on the silk-worms, and economize their food.

If this preliminary information is necessary, it may not be without advantage exactly to know what quantity of the mulberry leaf the silk-worm consumes in its four first ages.

For the quantity of food I fix, I must suppose the following circumstances to exist :—

That the silk-worms are kept until the first casting or moulting at 75° of temperature, between 73° and 75° until the second moulting; between 71° and 73° until the third; and lastly, between 68° and 71° till the fourth moulting.

One of the foundations of the art of rearing the silk-worm, is to know and determine the various degrees of heat in which, according to their ages, the silk-worms are to live; if this precept is not rigidly enforced, nothing can be performed with exactness *.

* The writer of an article upon silk-worms inserted in M. Rozier's Course of Agriculture, Paris edition, 1801, thus expresses himself relative to the heat suitable to silk-worms.

" It cannot be said that silk-worms are injured by any degree of heat in these climates, however considerable it may be. Native of Asia, it must needs be accustomed to heat more intense than it can experience in Europe ; but the sud-

The silk worms-proceeding from one ounce of eggs, consume—

den change from moderate heat to violent heat, it cannot bear. Rapid changes in general from heat to cold, and cold to heat, are highly injurious to it. In its native climate it is not exposed to these vicissitudes, and therefore thrives well, without requiring all the care we are obliged to bestow on it. With us, on the contrary, the temperature of the atmosphere is so variable, that without artificial means, we could not fix it in our laboratories, for rearing silk-worms. A series of experiments has proved, that in France, 68° is the most suitable to the silk-worms. Some cultivators have raised it as high as 73° and even 77° with good success. We must not lose sight of this fact, that it is not heat that affects the silk-worm, but sudden transitions from one temperature to another. Such as making it pass from 68° to 77° in one day, I am convinced would greatly annoy it, and injure its health. If it happen to be necessary to hasten the worms in consequence of the advanced state of the mulberry leaf, which cannot be retarded, it should be done gradually, so that they perceive not the alteration. The silk-worm suffers as much from difficulty of breathing in bad air, as from sudden changes of temperature. M. Boissier de Sauvagues will shew us by his experiments, to what degree the heat may be raised in rearing silk-worms, without fear of injuring them. "One year, when hurried by the early growth of the mulberry leaves, which were developed towards the latter end of April, I gave the silk-worms 100° of heat during the two first days after hatching, and about 95° during the remainder of the first and second age. There elapsed only nine days from the hatching, until the second moulting or casting, inclusively. Those of the trade who saw the process, could not imagine the silk-worms would be able to stand so intensely hot and overcoming an atmosphere. The walls and wicker hurdles were so heated, they could scarcely be touched. All thought they must be burnt, must perish; however, all went on perfectly well, and to their great surprise, I had a most abundant

E 3

1st. In the first age, (that is to say, when *all* are hatched, removed, and distributed upon the

crop. I afterwards tried giving the silk-worms, in their first age, from 93° to 95°, 89° to 91° in the second age ; and it is remarkable, that the duration of these two ages was nearly similar to that of the preceding experiment, in which they had experienced some degrees more of heat. Perhaps there may be a degree of heat beyond which we cannot affect the progress of the silk-worm. It is to be added, they had an equal proportion of food in both experiments to that which is given in the common manner of rearing silk-worms. It is singular, that these worms, thus hastened in their two first stages, consume only five days in moulting the third and fourth time, although with only a temperature of 82° : whilst those worms that have not been hastened, take seven or eight days for each of the two last moultings, in an exactly similar degree of temperature. It appears sufficient to have given the constitution of the insect an impetus to regulate the quick succession of its changes.

"This impetus, which we have been describing as operating such rapid growth, also gives the insect vigour and activity, which they preserve through their after ages, and prevents diseases ; thus the hastened, or forced cultivation, presents a double advantage. It also shortens the care and attendance necessary for silk-worms, and sooner ends the anxiety of the cultivator, who must necessarily feel anxiety until the cocoon is gathered.

"To follow this method, it is requisite well to observe the advancement of the season ; the shooting of the mulberry leaf, whether it is checked by cold : if again, the growth of the leaf is delayed, and that heat should soon after set in, and ripen it more quickly than was expected, as often occurs, it would be advantageous then to hasten the worms by heat ; for if they are allowed to delay from want of heat, their first age is prolonged, and the mulberry leaf will grow, and harden, and become unfit for them. The essential point is, that their progress should follow that of the mulberry leaf. If

sheets of paper, which includes at least two days)
(Chap. V. § 2.) six pounds of leaves, well sorted,
and chopped very small.

2d. In the second age, they consume 13 pounds
of sorted clean leaves, chopped rather more
coarsely than the food for the first age.

3d. In the third age, they consume 60 pounds
well sorted, and less chopped.

4. In the fourth age, 180 pounds well sorted,
and still less chopped than that of the third age.

Some circumstances may modify the proportions
specified above, but these variations are not important, supposing the cultivator to act with consi-
derate intelligence, and well to time the hatching of his silk-worms with the springing of the
young leaves, and then their growth with the progress of the leaf through the other stages of existence.

If the eggs be hatched before this favourable
time, they must be thrown away, and others should
be hatched, particularly if the unexpected badness of the season stops or delays the vegetation
of the mulberry, as it often has happened, and as

cultivators adopt this method, they must put the eggs to
hatch ten days *later* than they would require to be laid to
hatch in the ordinary way. And they must calculate the duration of the different ages of the worm, and so manage that
the completion of the rearing, or fourth age, should fall into
the time in which the leaf has attained its full growth."—
The Translator.

it happened in 1814. Whereas, by beginning to hatch the eggs when the season is fairly advanced, should it suddenly grow unfavourable, it is easy to gain some days, by delaying the rapid progress of the worm without any risk, as may be seen in the Table at the end of this work.

It may, however, happen that if the season continues so bad as to render the leaf sickly and weak, (Chap. III.) a greater quantity of the leaf will be required than I have here stated.

The quantity of leaves I have mentioned may be too plentiful, if the season prove so favourable as to make the leaf less watery and more nutritive.

When I fixed the proportion of leaves, I, of course, made my statement upon an average of the ordinary course of the seasons, which must always be understood when general rules are laid down.

The only case in which the quantity of leaves fixed, in these general rules, for the consumption of the worms, will be found unnecessarily excessive, will be if the worms have been ill attended, and fall sick, pine away, and that many die. The quantity of leaves necessary for silk-worms was determined, after having made the most exact experiments repeatedly. Taking for granted that the silk-worms are maintained in the degrees of temperature which I have indicated, and with the view, as much as possible,

to economize the leaves; because, when an exact
sufficiency of food is given to the worm, it eats
with greater relish, digests well, and is strong.
An object of great use in the art of rearing silk-
worms is to contrive to obtain the greatest possible quantity of fine cocoons, with the least
quantity of leaves. In managing upon this principle, the more leaves there] are the greater will
be the proportion of cocoons, and consequently
the greater the profit.

I do not exaggerate when I say, that in many
laboratories there is a quarter, nay, a third more
of the leaf consumed than is required; which is
not only a waste of leaves, but is the origin of
many inconveniences which assail the silk-worm,
as we shall shew hereafter. The cares which the
silk-worm requires in its four first ages are neither
numerous nor puzzling; although it is in those
ages, and particularly in the two first, that the
strength of their constitution is formed, upon
which the ultimate success depends.

The silk-worm seems doomed by its natural
constitution to have but few days of vigour from
its coming forth, until after its fourth moulting.
It is only healthy in the periods between the
moultings. The two first days after it has cast
its skin it eats sparingly; it then becomes voracious; this hunger soon diminishes, and even
ceases. These phenomena occur in every moult-

ing. Thus, from the miserable condition of this
insect, notwithstanding the strength of its constitution, if it is not treated with infinite care at those
times when it requires care, it suffers, sickens,
and dies.

It is on this account that I have thought it
might be useful to give in this and the following
chapter, a diary of the care to be taken of silk-
worms, that it may be known what is to be done
for them day by day.

I must, however, previously, make a few general remarks on the enormous difference, in result, which real care produces.

I do not here mean to allude to the slight and
partial differences, that may be considered as only
exceptions and accidental. In such cases the intelligent cultivator, well informed by the precepts I am going to lay down in this chapter, will
easily know, if he is attentive, how to proceed to
prevent inconveniences, and how to remedy them.
I mean to speak here of those differences which
are caused by ignorant and ill-directed management.

Hitherto it has been generally thought, in
quoting facts and experiments, that, whatever
were the quantity of eggs intended for a laboratory, the quantity of cocoons never bore any proportion to the eggs; and that, on the contrary,
the greater the quantity of eggs, the less the

quantity of cocoons ; and thus the observation is, that if five ounces of eggs, produced thirty-five pounds of cocoons per ounce ; four ounces would produce forty pounds per ounce, three ounces forty-five pounds per ounce, two ounces fifty pounds, &c. &c.

These differences, it should be known, do not depend on the organization or natural condition of the silk-worm, but are solely to be ascribed to error and ignorance. Facts and most evident reason surely prove that, if the silk-worm have had space, if the degrees of temperature have been exactly regulated, if the necessary quantity and quality of food has been given the silk-worms, and that all the care I have recommended has been practised, the quantity of cocoons should be, and always will be, proportioned to the quantity of eggs that were hatched. Those who do not obtain this result should attribute their failure to the erroneous system they have adopted. My laboratories are of various sizes ; that which I am going to describe is calculated for the reception of the worms proceeding from five ounces of eggs ; the other laboratories equally yield cocoons in proportion to the eggs which I have hatched.

I should allow the advantage of my manner of rearing silk-worms to be most trifling, if it were only in the product of the hundred and ten, or

E 6

hundred and twenty, pounds of cocoons from each ounce of eggs, which others obtain, consuming the same quantity of leaves, and differing only in the hatching of two ounces of eggs. But as I said before, the great and principal aim of the art of rearing silk-worms, is to obtain from one given quantity of mulberry-leaves the greatest possible number of cocoons of the finest quality. It is not the trifling loss of an ounce of eggs which should induce a change of system of habits, but the following advantages: for it is a true fact, that,

1. When with one ounce of eggs one hundred and ten or one hundred and twenty pounds of cocoons are obtained, about one thousand six hundred and fifty pounds of the mulberry-leaf will be used. (Chap. XIV.)

2. That when only fifty-five or sixty pounds of cocoons are produced from one ounce of eggs, about one thousand and fifty pounds of mulberry-leaves have been used; under this supposition, it would appear that two thousand one hundred pounds of leaves are requisite to produce one hundred and ten, or one hundred and twenty pounds of cocoons.

3. That one hundred and ten, or one hundred and twenty pounds of cocoons, obtained from one ounce of eggs, are worth a great deal more than a similar quantity obtained from two ounces of

eggs. It is easy to prove these facts rational. I stated (Chap. V. § 3.), that thirty-one thousand one hundred and sixty-eight eggs, which constitute an ounce, might produce about one hundred and sixty-five pounds of cocoons. If, upon this statement, we consider the loss of worms to be great when we obtain one hundred and twenty pounds of cocoons from one ounce of eggs, the loss will be essentially greater, should we only obtain sixty pounds. It is natural that by this greater mortality a greater consumption of leaves should result, as the worms, which do not reach the consummation of the cocoon, feed more or less as well as those that accomplish the cocoon.

The great mortality of the worms must also affect the quality of the cocoon ; for who could suppose that two-thirds of the worms, proceeding from one ounce of eggs, should die, without some error or want of care ? If want of care, or erroneous care, should have caused the death of such numbers, are we not justified in thinking that part of those that remain may be weakened and injured?

This would be still more forcible, if, as it frequently happens, the sixty pounds of cocoons should be reduced to forty-five, thirty, fifteen, &c. &c.

Whereas, if one ounce of eggs shall have produced, by the means I have stated, one hundred and twenty pounds of cocoons, they will be fine, and will sell well; three hundred and sixty, at most, will produce a pound and a half; and eleven or twelve ounces, at most, of these cocoons will yield an ounce of exquisitely fine silk, as I shall, hereafter, demonstrate. When only fifty or sixty pounds of cocoons come from an ounce of eggs, it may generally be presumed that they are of inferior quality to the above, that they are not so valuable, and it will require four hundred at least to make one pound and a half; and above thirteen ounces of these cocoons, instead of eleven or twelve ounces, to form one ounce of silk.

Moreover, when the worms have not been properly managed, there is no certainty as to the quantity of the cocoons that will be gathered; and it happens continually, that the same cultivator will, from the same quantity of eggs, and the same quality of the leaves, obtain at one time a number of cocoons, at another time few, and sometimes none.

It would be interesting, as well to the government as to individuals, to compare the quantity and quality of the cocoons produced upon my plan, with those produced upon the system which is generally adopted ; to establish, by reason and facts, which is the most rational and profitable method. If it were calculated afterwards, the loss occasioned by ignorance every year, and particularly what was lost in 1814, the immensity of the value of these losses would doubtless cause astonishment. (Chap. XV.)

This chapter shall be divided into four paragraphs :—

1. Rearing of the hatched worms until the end of the first age.

2. Rearing of the worms in the second age.

3. Rearing of the worms in the third age.

4. Rearing of the worms in the fourth age.

1. *Rearing of the Worms in the first Age.*

We left in the small laboratory the worms hatched from the eggs at 75° temperature, and distributed upon sheets of paper, (Chapter V. § 2.), in small squares of about ten inches a side.

Let us now begin their training. Supposing it is required to rear five ounces, which forms a good-sized laboratory. The space and quantity of leaves must be proportioned to the stated number of silk-worms. Having chosen a large laboratory, I had in view to shew that all things being equal, the practical results upon a large scale are equally applicable to the small scale, and are always similar.

First Day's Training.—When the worms proceeding from five ounces of eggs have accomplished their first casting of skin, they should occupy a space of nearly thirty-six feet eight inches square; thus the sheets of paper containing the worms, should be put upon wicker tables or trays of such dimensions.

The first day after the coming forth, and the distribution of the silk-worms, they should be given, in four meals, about three pounds three-quarters of single soft leaves, chopped very small, dividing the time, so as to allow six hours between each meal, giving the smallest quantity for the first feeding, and gradually increasing the quantity at each meal.

It is very beneficial to chop the leaf very small during the first age, and to scatter it lightly over the worms. The more the leaf is chopped, the more fresh-cut edges are there to which the young insects fasten themselves. In this manner a few ounces of leaves will present so many edges and sides, that two hundred thousand insects may feed in a very small space. In this state they bite the leaf quickly, and it is consumed before it can be withered.

A quantity of leaves, ten or twenty times more abundant, that is not chopped small, would not be sufficient for this quantity of worms, because they require to find at once, and in a small space, the means of feeding easily.

If care is not taken to chop the leaf small, and to give the young worms space when they are

small, a great number must perish, as they will
contract diseases, and lose the equality they ex-
hibited. (Chap. XII.) The worm that cannot
eat dwindles, becomes extenuated, weak, sickly,
and perishes under the leaf. This object, which
appears trifling in itself, is, however, of great
importance, and deserves unceasing attention.
To chop the leaf for the different ages of the silk-
worm, I make use of knives, and a variety of
sharp tools. (Figs. 14, 15, 16.)

I feed the worms regularly four times a day ;
and I manage so as never to give the whole quantity at once, as stated above, because after the
distribution of each meal it is better to observe
if some food should not be added in different
spots. It is sometimes good to give them a little
food at intermediate times, as will be seen hereafter.

The quantity of food I have fixed, and which
I shall again specify, is that necessary for the
whole day. In about an hour and a half the silk-
worm devours its portion of the leaves, and then
remains more or less quiet. Whenever the food
is given, care should be taken gently to spread
and widen the small squares by degrees. If any
of the chopped leaf should be scattered, it may
be swept with a small broom into its place again.
(Fig. 17.)

Second Day.—On this day, about six pounds

will be needed, chopped very small. This will suffice for the four regular meals, the first of which should be the least, increasing them as they proceed, as was done in the meals of the first day.

The worm now begins to change in appearance ; it no longer looks so dingy, or so bristled ; the head enlarges, and whitens considerably.

The squares should be spread and enlarged every time the worms are fed, to make room for them.

Third Day.—This day, twelve pounds of soft leaves, chopped very small, will be required for the four meals; the worms will now feed with avidity, and nearly the two-thirds of the sheet of paper should be engrossed by them.

To satisfy the increased hunger of the insects, they should be given a pound and a half of chopped leaves, slightly scattered over them : should they devour it quickly, in an hour, it would not be advisable to wait the five hours to give the second meal, but they should have an intermediate feeding, about half the quantity of the first meal, scattering the leaf very sparingly over them. I do not here fix the number of ounces of these intermediate meals, because it could not be done with exactness ; it must be more or less calculated upon the quantity of leaves that is to be given them in the course of the day, and the

disposition of the worms. This day the head of
the silk-worm is much whiter ; the insects have
perceptibly grown larger; scarcely can any hairiness be perceived on them with the naked eye.
The skin is of a sort of hazel colour. When seen
through a magnifying glass, their surface looks
shining, and their head is of a silvery bright appearance, like mother-of-pearl, and rather transparent.

Fourth Day.—This day, six pounds twelve
ounces of chopped leaves should be given, for the
quantity should be diminished as the appetite decreases; the first meal should be of about two
pounds four ounces, and the other meals should
decrease in proportion as the quantity of leaves
given before appears not to have been thoroughly
eaten.

The cultivator must regulate the intermediate
meals, upon the apparent appetite of the silk-
worms, taking the food for them from the quantity of leaves allotted for the whole consumption
of the day. The space on the sheet of paper
must visibly get covered with the worms. It is
important, in this first age, to give the worms
plenty of room, by gently separating and spreading them, to avoid as much as possible their sleep-
ing in heaps together.

The constant care of enlarging the squares by
degrees when the worms are fed, will gradually

lead them to stretch out as they grow, and prevent their getting into heaps, which is very injurious to their constitution, health, and to that
equality in size, which it is so desirable to maintain among them.

At the beginning of this day many of the silk-
worms begin shaking their heads, which indicates
that they feel over-loaded by their covering, or
skins. Some of them eat little, but keep their
heads reared up; with a magnifying glass it may
be seen that their head is increased much, and
grown very shining. The whole body of the insect seems transparent, and those that are near
their time of moulting, when seen against the light,
are of a yellow livid tinge ; towards the close of
this day, the greatest number of the silk-worms
appear torpid, and eat no more.

Fifth Day.—This day one pound and a half of
young leaves, chopped small, will be about sufficient. They should be scattered very lightly several times in the day on the spots, and on the
sheets of papers, where there appears still to be
worms feeding. If the quantity should not be sufficient, more may be added ; as also, should the
worms have left off feeding, it would be unnecessary
to distribute any further quantity. What I have
said as to the different variations of quantity required by the silk-worms of this age, is applicable
to all the other ages. I cannot sufficiently urge

economy and regularity in the distribution of the leaves.

Towards the end of this day the worms are torpid, and a few begin to revive even.

After the first moulting, the silk-worm is of a dark ash-colour, shewing a very distinct vermicular motion ;. the rings that compose its body stretch and shrink more freely than heretofore.

I must here repeat that it is of the utmost import that the food should be chopped very finely, first with a knife, and then with the double-bladed hashing tool, which may be seen (Fig. 15.)

When the weather admits of it, the leaves should be gathered several hours before the meal is given ; they last very well a day, and more if kept in a damp cool place, where there is no draught of air. It is always desirable the leaf should have lost its first sharpness, and not be given to the worms till six or eight hours after it has been gathered.

I shall now draw up a general view of this paragraph, and add a few observations that appear to me useful.

The first age of the silk-worm, reared in the temperature I have indicated, is almost always accomplished in five days, exclusive of the two days employed in their coming forth, and being removed and distributed.

In this first stage, the silk-worms proceeding from five ounces of eggs, have consumed thirty

pounds of picked leaves, chopped small; in adding four pounds and a half, the refuse picked off the
leaves, the weight will make 34¼ pounds of mulberry leaves, or about seven pounds of leaves
from the tree to each ounce of silk-worms.

To complete the exactness of these observations,
we must add two other alterations, to which the
silk-worms are subject before their moulting.

1st. We have seen (Chap. V. § 4.) that, to form
an ounce of silk-worms just hatched, it requires
54,626 worms; after the first moulting, 3840 are
sufficient to make up that weight. Thus the silk-
worm has increased, in about six days, fourteen
times its own weight.

2d. Before the above six days, the silk-worm was
about a line in length, and after those days it is
about four lines long.

In the first age the air of the laboratory should
only be renewed by opening the doors. The necessary degree of temperature must be maintained
by the stove, and wood-fires in the fire-places, as
we shall shew hereafter.

Nothing further is necessary for the thriving of
the worms, and their healthy continuance.

2. *Rearing of the Silk-worm in the Second Age.*

Nearly seventy-three feet four inches square of
table or wicker-trays are needed for the accommodation of the worms proceeding from five ounces

of eggs, until the accomplishment of their second casting, or moulting.

These trays or tables, as I before mentioned, should always be covered with strong paper. The temperature to keep the worms in, during their second age, should be, as I said before, 73° and 75°. The insects should not be lifted from their litter until they are nearly all revived. I will explain the manner of removing them shortly. There is no harm in waiting till they are all well awake and stirring, even should it be for twenty or thirty hours from the time when the few first began to revive.

When a great number of worms issue from the sheets of paper where they were placed, it is a sign that they should be removed from their litter, and removing them a little sooner the others will soon revive also.

We have said already that, during the first age, most cultivators destroy the life and destroy the health of a vast number of worms by not attending to them sufficiently; and consequently it frequently happens that beyond that age they have worms of unequal sizes, a very great defect, which is never remedied.

This inequality, and the evils resulting from it, as I shall demonstrate (Chap. XII.), are caused, 1st. By not having placed the silk-worms in a space proportional to their growth in the course of their first age ; which has allowed of some having fed well, while others could not feed ; of some remaining under the litter, others upon it, which latter had the benefit of free air, instead of a close mephitical atmosphere ; some began to fall into their torpid slumber sooner than others, and being under the leaves, have moulted the last ; others, in short, became torpid latest and revived first, because they were upon the surface of the leaves, unloaded and unoppressed.

2d. By not having placed the sheets of silk-worms hatched the first day in the coolest parts of the laboratory.

3d. By not having placed the latest hatched worms in the hottest part of the laboratory. (Chap. IV. § 3.)

4th. And finally, by not having given the last-hatched silk-worms intermediate meals, to bring on their growth a little faster.

It follows from this want of attention, that when the silk-worms should pass from their first casting, or moulting, to the second, some worms are torpid, some are reviving and beginning to feed, and some have not yet fallen into the torpor which is to precede their change ; and thus on one wicker-hurdle may we see silk-worms of all sizes, which is most troublesome, to say no more ; there is besides a great chance of the smaller worms perishing in their progress.

All these losses will be avoided by strictly following the rules I have given. It is then useful
to wait the revivification of the greatest number
of the silk-worms before they are fed, particularly
as these insects, when they have cast their skins,
need free air and gentle heat more than food.

Their organs assume greater consistency by exposure to the air. The small scaly muzzle, which
they lose in moulting, is replaced by one softer,
which the air indurates ; and till the small jaws, or
sawing teeth, have acquired hardness and strength,
they cannot divide the leaves. It is easy, with a
magnifying glass, to observe the labour and effort
with which the silk-worm endeavours to cut and
gnaw the leaf. Having given those general views
which I thought might be useful, we will now resume our diary.

First Day of the Second Age.
(Sixth of the Rearing of the Silk-worm.)

For this day will be needed nine pounds of
young tender shoots, and nine pounds of mulberry
leaves, well picked and chopped small.

The space of seventy-three feet four inches
square of tables or hurdles required for the second
age of the silk-worms proceeding from five ounces
of eggs, should be duly prepared ; and when
nearly all the worms are roused, and begin moving their heads, and rearing up as if they sought

F

something, those at the edge of the paper having already left the litter on whieh they had lain, preparation should be made to remove them, that the sheets of paper may be cleansed.   The worms should be removed from those sheets of paper first where they are perceived to be most revived and stirring.   Small twigs of the young shoots of the mulberry tree, with six or eight leaves on them, should be put over the silk-worms ; these boughs should be placed so that, when spread out, there may be an inch or two between them.   When one of the sheets of paper is thus covered with silk-worms, another is begun, and so on till all are completed ; this must be done speedily.   There should be some boughs left, which will be wanted. These boughs will gradually be entirely and thickly covered with the worms.   The small portable tray should be ready (Fig. 9.) upon which the boughs covered with worms must be put quickly, when taken off the sheets of paper.

Instead of forming small squares, as was done for disposing of the new-hatched worms, long strips should be laid down the middle of the wicker hurdles, prepared, so that by widening them on each side, when arrived at the consummation of the second age, the whole space of seventy-three feet four inches of the hurdle should be entirely covered by the silk-worms.

The use of the small portable trays is beneficial ;

they carry and place with ease the small boughs loaded with silk-worms, and by inclining them obliquely upon the hurdle, slip off the boughs gently into the strips allotted for them, being careful, lightly with the hand to move those that may not be properly placed, filling the vacant spots with them, so as to render the distribution regular.

This operation concluded, some worms will be found to have remained upon the litter; with fresh boughs these may be removed as the others were, and distributed upon the hurdles: should any after this remain torpid in the litter they may be cast away.

The leaf on which the worms were removed affords them a slight meal; they devour and entirely pierce it, until the fibrous part alone is left, which shews that the mere contact of good warm air has sufficed to give strength and hardness to the jaws of these insects, which they possessed not before.

It is observable that the silk-worms like the tender boughs so much, that they are found heaped upon them, even when they have entirely eaten the leaves off, and never leave them to return to the litter below. This remark will doubtless do away the idea which is prevalent, that the silk-worm likes the litter to lie on, and to feed upon it.

The means I have described for removing the

silk-worms is the best, and suited to all the different ages.

The worms thus removed upon clean trays, with fresh boughs, get strong and revived.

An hour or two after the worms have been placed upon the hurdles, they should be given a meal of three pounds of leaves chopped small.

When the boughs are stripped of the leaves by the worms, there will be bare spaces in the paper, and the boughs swarming with worms. To remedy this, the leaves should be gently laid on those bare places, and the worms stretching upon them, will equally spread and fill the strips. The space occupied by the worms should be widened a little when they are first fed. It should be remembered to sweep up with the small broom the leaves that may be scattered. I must here state that the care or attention I have here recommended, as well as in other parts of this work, and which I myself practise for the success of my silk-worms, is neither tedious nor difficult, although it may at first appear complicated.

In the remainder of this day, the silk-worms should have, in two meals, the remaining six pounds of chopped leaves, with an interval of six hours between each, or according to the hours of the day which remain.

When the silk-worms have been removed to the clean hurdles, those they have left should be

thoroughly cleaned, the sheets of paper cleansed
and rolled up, and taken out of the laboratory.

If the substances remaining on the paper are
examined, they will be found to consist of a mass
of the fragments of leaves, and moist excrement,
without any unpleasant effluvia, the weight of
about seven pounds and a half.

From the first day of the rearing of silk-worms
until the first moulting they have consumed thirty
pounds of leaves; twenty-two pounds eight ounces
of substance have contributed to the growth of
these insects, or have evaporated. In the first
age the worms void a small portion of excrement,
resembling fine black powder; of the seven
pounds eight ounces of substance there are only
about ten ounces of excremental matter.

Second Day of the Second Age.

(Seventh of the Rearing of the Silk-worm.)

This day will be required about thirty pounds
of chopped leaves; this quantity, divided into
four portions, should be given at intervals of six
hours; the two first meals less plentiful than the
two remaining. It is very necessary gradually to
widen on both sides the strips in which the worms
are distributed, that at the close of the day two-
thirds of the allotted space should be covered.

The body of the worm now acquires a clear
hue; the head enlarges and becomes whiter.

F 3

Should some places be thinly covered with worms, by placing small boughs where the worms lie thick they will fasten on them, and may then be removed to fill up the places which were not sufficiently covered ; the equality of the worms being very desirable, it should be greatly attended to, and those means practised which I have stated through all the moultings, and whenever circumstances seem to require them.

Third Day of the Second Age.
(Eighth of the Rearing of the Silk-worm.)

This day thirty-three pounds of chopped leaves well picked will be necessary, and this time the two first meals should be the largest.   The leaves should be distributed in proportion as they are wanted, distributing them with attention, because the voracity of the silk-worm abates towards evening ; and many worms shew, by rearing their heads, and not eating, that they are approaching the period of torpor, and some already are become torpid.

The strips should continue to be widened, so that at least four-fifths of the hurdle should be covered.

Fourth Day of the Second Age.
(Ninth of the Rearing of the Silk-worm.)

This day only nine pounds of picked leaves, chopped small, will be required, which should be distributed in the same manner as before, as may be wanted, lightly and carefully scattered over the worms.

This day the silk-worms sink into torpor, so that the next day they will have cast their skins and will be roused, and thus will the second age be accomplished.

Let us now generalize this paragraph, and add our observations to the result.

In the four days which the second age has lasted, the silk-worms, proceeding from five ounces of eggs, have consumed 90 pounds of picked leaves chopped small, including nine pounds of the young shoots of the mulberry; adding to this about 15 pounds of refuse and pickings of the leaves, we shall have drawn about 105 pounds' weight from the tree, and given it in a proportion of 21 pounds to each ounce of silk-worms.

The alterations which the silk-worm undergoes, besides that of the moulting in the second age, are as follows :—

Their colour is become of a light grey, the hair is hardly to be perceived by the naked eye, and is become shorter ; the muzzle, which in the first age was very black, hard, and scaly, became immediately, upon moulting, white and soft, but afterwards again grew black, shining, and shelly, as before.

F 4

As the insect grows older, at each moulting, its muzzle hardens, because it needs to saw and bite larger and older leaves.

There appear now two curved lines opposite each other, upon the silk-worm's back.

The length of the silk-worms, in the first age, was rather less than four lines; in the second age, of rather more than six lines.

In four days it has increased its average weight fourfold; when issuing from the first moulting, 3240 silk-worms formed one ounce, at this period 610 will form this weight.

As the insect grows, it breathes more freely, its excrements are more plentiful, which, as the number of hurdles also increase in the laboratory, makes it necessary the interior air should be more renovated in the laboratory; and to effect this, the ventilator in the floor, and the aperture made in the door, should be opened. (Fig. 18).

Should there be no wind or cold in the atmosphere, the ventilator may be left open until the thermometer has lost a degree, or, indeed, two complete degrees.

Then all should be closed; the temperature again rises, and thus has the interior air been thoroughly renewed and purified.

3. *Education of the Third Age.—First Day of the Third Age.*

(Tenth Day of the Rearing of the Silk-worm.)

In this first day, 15 pounds of the small shoots will be necessary, and equally as much of the picked leaves, chopped rather less than it has hitherto been, and at the close of the age may be still more coarsely chopped.

At this age, the worms of five ounces of eggs should occupy nearly 174 feet square of space. and, consequently, the space in hurdles or tables must have been prepared and covered with paper for the silk-worms.

The temperature of the laboratory during this third age should be from 71° to 73°.

The worms that have accomplished the second age should not be removed from the wicker hurdles, until they are all nearly roused; part will rouse the ninth day, part the tenth. There would be no harm if those first revived should wait 24 hours, till the rest are all roused.

It is very easy to know the worms that are revived in this age ; they issue from their old skin with so different an aspect, that anybody may distinguish them without the aid of description.

A never-failing sign that the silk-worms are roused, is an undulating motion they make with their head, when horizontally blown upon.

The impression of the air thus forcibly blown over them is disagreeable and painful to them, when they have newly cast their skins; but gentle motion of the air through the laboratory is pleasant to them, and does them good, provided the renewed air is not colder than their usual atmosphere.

They should be removed in the same order and manner as in their former age (§ 11.)

The space of 174 feet, allotted to the third age, should be disposed in a strip down the centre of the wicker hurdle, and of nearly half the width of the hurdle, so as to leave rather more than a quarter's width down each side of the strip.

When the space is well ascertained which the silk-worms are to occupy in their different ages; there is nothing more easy, more useful, and more economical, than to remove, cleanse, and place them in the manner I have described.

Once placed upon their wicker hurdles, they are no more touched until their casting is accomplished; they feed well without interfering with one another, and without requiring to have the intervals on the sheet of paper cleaned. Their litter does not become mouldy, unless there should be a very unusual and continued dampness of weather, but is of a fresh green colour, thinly scattered, nearly dry, and composed of the fibres of the leaf, and little portions of the leaf fallen from the mouth of the animal: instead of disgusting those that attend them, they are pleasing to look at.

The fifteen pounds of young shoots afford the silk-worm its first meal, as in the preceding age.

When the silk-worms have eaten the leaves upon the shoots, they should have a second meal of about seven pounds and a half of the leaves, carefully filling with leaves the space between the shoots, to equalize the distribution of the worms upon the strips.

I must unceasingly repeat it, to ensure the silk-worms continuing of an equal size, the cultivator must always watch those persons who distribute the food, that it may be perfectly even, and all the worms enabled to partake of it. A waste of the leaves is not only a real loss, but also is apt, by thickening the litter in heaps, to ferment much more speedily than the more fibrous litter could have done, and thereby cause disease.

The worms should have their last meal this day of seven pounds and a half of leaves, which completes the feeding of the day.

Should the removal of the litter be late in the day, so that the silk-worms could not be given the three meals, that portion of the leaves may be added to those of the following day.

Two active, handy persons should take but one

hour to remove and distribute the worms upon
the 174 feet of hurdles.

As fast as the worms are lifted off, the litter
should be carried out of the laboratory, rolled
up in the sheets of paper ; when taken out, the
litter should be examined, in case any of the
torpid worms may be remaining in it, and if this
is done in any place sheltered from rain and wind,
far from injuring the worms, it will tend to rouse
them sooner than they would have roused in the
laboratory, to which they must be taken back by
offering them young shoots, to which they will
fasten, and thus be carried safely.

The latest worms should be placed apart, as
their next moulting will be a day later also, or
should it be desired to bring them on equally with
the others, by giving them rather more space
between them on the hurdles, and putting them
in the hottest part of the laboratory, this may be
managed.

Now, as the worms begin to eat more, it is
useful to employ the square basket (Fig. 19.),
with which twice the work may be done in feeding the worms, compared to the usual method
of holding the shoots and leaves in an apron,
and feeding the worms with one hand only ; by
means of this basket, which may be suspended
with a hook, and slide in a groove along the edge
of the wicker trays, the feeder may arrange and

distribute the food with both hands, and thus feed two trays of worms at once by standing on high steps or ladders. (Figs. 20, 21.)

The weight of all the litter of the second age is of about 21 pounds. If the leaves have been thoroughly eaten, the dark excrement weighs alone about six pounds. It must be observed, however, that the degree of moisture of this substance makes the weight differ.

As from the close of the first age to the accomplishment of the second, 90 pounds of mulberry leaves have been distributed upon the hurdles, it is evident that 69 pounds of leaves have nourished these insects, and have exhaled in evaporation.

After two or three meals this day, there is a very sensible change in the silk-worms. They are much larger, their muzzle is grown longer, and their colour clearer.

Second Day of the Third Age.

(Eleventh of the Rearing of the Silk-worm.)

This day 90 pounds of picked leaves, chopped, will be needed.

The two first meals the least copious, because towards the close of the day the silk-worms grow voraciously hungry.

The strips should be widened whenever they are fed, to allow them room.

Third Day of the Third Age.

(Twelfth of the Rearing of the Silk-worm.)

This day there should be given 97 pounds of picked leaves, chopped, divided in four meals, the two first meals the most plentiful. Towards evening the hunger of the silk-worms decreases, consequently, the last should be the least meal.

This day the silk-worms grow fast, their skins whiten, the bodies are nearly transparent, and the heads are longer.

If a hurdle of worms is seen against the light, before the worms are fed, they seem of the colour of whitish amber, and appear powdery.

The contortions they begin to make with their heads, shew that their change approaches.

Fourth Day of the Third Age.

(Thirteenth of the Rearing of the Silk-worm.)

This day about 52 pounds and a half of chopped leaves will be sufficient. The decrease of food is consequent upon the diminution of appetite already mentioned; many of the worms are already torpid.

They should be given four meals, the largest first; and the last, the least meal. Those only that seem to require it should be fed.

Should a great number of silk-worms, on one table, be torpid, whilst others continue to require food, these should be given a slight meal without waiting for the stated hour of their feeding, to satisfy them, that they may sink into torpor quickly; this care is of consequence, and these intermediate meals are very beneficial.

Fifth Day of the Third Age.
(Fourteenth of the Rearing of the Silk-worm.)

This day 27 pounds of picked leaves, chopped, will do, which must be distributed as they are needed: this will be about the quantity; if it is not enough, more may be added; if too much, less given.

These two last days the silk-worms begin to cast about some silk down.

The insect seeks free space to slumber in, dry and solitary spots, rearing its head upwards, which is known by finding it on the edges of the paper, where any stalks stick up, upon which it retires. All of them not being able thus to separate from each other, are obliged to remain upon their litter, but testify uneasiness by rearing up their heads.

When on the point of sinking into torpor, they completely void all excremental matter, and there remains in their intestinal tube, a yellow-ish lymph alone, rather transparent, and which supplies almost all the animal fluid in them. And this it is that, before the surface of the skin they are going to cast becomes wrinkled and dry, causes them to appear of a yellowish white colour like amber, and semi-transparent.

When the worms prepare for the third, and even fourth, moulting, the air of the laboratory should be gently agitated, but the temperature should not be much varied; this may be done by more or less opening the ventilators in the ceiling, and those in the floor, constructed as we shall hereafter explain. (Chap. XIII.)

Sixth Day of the Third Age.

(Fifteenth of the Rearing of the Silk-worm.)

On this day the silk-worms begin to rouse, and thus accomplish the third age.

The general view of this age presents the following result.

In six days the silk-worm goes through its third age.

In this age, those worms proceeding from five ounces of eggs have consumed nearly 300 pounds of leaves and young shoots ; adding to this weight 45 pounds of refuse and pickings, 345 pounds' weight have been drawn from the tree, or 69 pounds given to an ounce of eggs.

The muzzle of the silk-worm, during the third age, has maintained a reddish ash-colour, and is no longer shining and black, as it appears in the two first ages, but it is lengthened and more prominent.

The head and body is much enlarged since the casting of the skin, even before they have eaten at all; proving that they were straightened in the skin they have cast, and being now unconfined, the air alone has expanded their bulk.

This growth, which is considerable, is more visible in this age than in the preceding.

When this age is completed, the body of the silk-worm is more wrinkled, particularly about the head ; they are of a yellowish white, or rather fawn colour; and to the naked eye they have no appearance of hairiness.

The membranaceous feet, and particularly those at the hinder extremities, have acquired much strength, and an adhesive quality which enables the silk-worm strongly to retain any thing it touches. In this third age we first hear, when the worms are fed, a little hissing noise, similar to that of green wood burning.

This noise does not proceed from the action of the jaws, but from the motion of the feet, which they are continually moving ; this noise is such, that in a large laboratory it sounds like a soft shower of rain; by degrees, when the worms fasten to their food, the noise ceases.

The average length of the silk-worms, which was six lines after the second moulting, is become, in less than seven days, above twelve lines.

The weight of the insect has increased fourfold in the same period; after the second moulting 610 worms made an ounce, now 144 only will complete that weight.

It has been sufficient during this age to open the ventilator, the door, and even the windows, when the weather was still and fine, so as to lower the temperature by a degree only.

In the damp close days, a light wood-fire, in the fire-place, renews the air by drawing a current, without fear of injuring the interior atmosphere.

During this age, it never happened to me to experience the exterior temperature, although higher than the interior, to be beyond the prescribed limits.

4. *Rearing of the Silk-worm in the Fourth Age.*

In this age, the worms proceeding from five ounces of eggs should occupy a space of about 412 feet square, which should be distributed in a manner similar to that practised in the former age.

The temperature of the laboratory should be from 68° to 71°.

In this fourth age, as in the fifth, there will
probably be days in which it will not be possible
to maintain the temperature of 71°, because of
the heat of the weather, as the season advances;
and in spite of artificial means, it may very probably rise to 73° or upwards.

This augmentation of temperature need create
no anxiety, because it does no harm. It is sufficient to prevent the circulation of air being
interrupted. The moment it is perceived that
the exterior atmosphere begins to heat the laboratory, the ventilators should be opened as well
as all apertures on the side unexposed to the sun.
I have seen, in the space of two hours, some of
my laboratories rise from 71° to 80°. I then
opened all the apertures, and the air being stagnant, I had some faggot-wood burnt in the fire-
place, placed in the angles (Chap. XIII.) to establish a complete current of air, and thus change
the air of all the rooms thoroughly. If, instead
of thus acting, when the heat of the season ceases
suddenly, (which augments the fermentation of
the litter) we should exclude the exterior air from
the laboratory, we may chance to lose whole
broods of silk-worms, because as they grow, the
mass of leaves and litter increasing, the dampness
proceeding from it will more quickly produce
fermentation, the heat would also augment, and

the air would soon be not only moist but pestilential. (Chap. XII.)

As we before said, the silk-worms should not
be lifted off the hurdles, after they have completed their third age, until they are nearly all
well roused, because, should the first-roused have
to wait a day, or a day and a half, it will not
hurt them.   Those early-roused should be put
in the coolest part of the laboratory, and the
late-roused worms in the warmest part.   If this
should be troublesome, it may suffice to give the
latest-roused worms more space by keeping them
farther asunder, and they will soon come up to
the others.

It is easy to tell by the thermometer which are
constantly the hottest parts of the laboratory.
And this knowledge will serve to render all the
silk-worms even-sized, particularly if those who
attend them have any practical skill.   All this
care is indispensable, if the worms are required to
draw their silk equally, and at the same period,
particularly as there accrue great evils, which I
shall hereafter demonstrate, when some of the
silk worms rise too much above the others. (Chap.
VIII. § 5.)

It is after the third moulting that the silk-
worms should be moved into the large laboratory,
in which they are to remain until the end.   The

space of this large laboratory should contain at
least 917 feet square of wicker hurdle or table.
Experience constantly demonstrates the advantage
of having buildings proportionate to what is required of them; as much on account of economy
of fuel, if the season were cold, as the convenience
of attendance. There would certainly be no great
objection, should there be two or three small
contiguous buildings, instead of one large laboratory, so that they afforded an equal space.

The only advantage that would thus be lost,
would be the great facility enjoyed in a spacious
building, of establishing, and maintaining, as we
shall shew, constant and regular currents of
air. (Chap. XIII.)

When we have the use of a single space, large
enough to contain the 917 feet square of hurdles
necessary for the accommodation of the silk-
worms proceeding from five ounces of eggs, it is
beneficial to choose the most convenient part of
the space, to place in it the 458 feet six inches
square of wicker hurdles upon which the insects
are to be deposited, until the accomplishment of
the fourth age, afterwards to distribute them
upon the whole space of 917 square feet.

There is nothing so easy for those whose laboratories are well regulated, as to determine how
the silk-worms are to be distributed in the 458
feet six inches, on coming forth from their third

age. It is only necessary to note on each hurdle its dimensions and the number of square feet; by which means in a moment may be seen which are the hurdles which must be used for this age, as well as for the preceding ages.

I must here repeat how advantageous to the art of rearing silk-worms is the practice of distributing them in regular strips and squares, which should be extended, and widened, and gradually filled with these insects, as they accomplish their various ages, and particularly as the hurdles are not cleaned in the fourth age, the litter, that spreads by degrees, not heating or contracting any effluvia, and not rising much.

2d. Because the leaf distributed upon evenly portioned spaces is entirely eaten before it is withered and spoilt.

3d. Because, by this practice, the worms can feed with facility, move with ease, and breathe more freely, all decisive advantages for these insects. (Chap. XIII.)

He must forego these advantages when the worms lie too thick; in that condition they cover the surface so closely, that the leaves on which they lie are wasted, as they cannot possibly eat them; while, on the contrary, when they have plenty of room, they seek, in moving, every atom of the leaf, and eat it up. Besides, when straightened, the action of their breathing-tubes is hindered and confined by the pressure, either superior or lateral, of one worm against the other ; whilst, when in full space, the action of their respiratory organs is free, which materially contributes to their health. (Chap. XII.)

First Day of the Fourth Age.
(Sixteenth of the Rearing of the Silk-worm.)

On this day 37 pounds and a half of the young shoots will be needed, and 60 pounds of picked leaves, coarsely chopped with the large blade (Fig. 16.)

When the moment of removing the worms from the hurdles comes, one or two hurdles only at a time should be covered over with young shoots. These shoots, loaded with worms, are afterwards put upon the portable trays and removed, as in the first moultings. Should there not be a sufficiency of small boughs, branches of 15 or 20 leaves tied together by the stalks, will answer the purpose.

The stiffer these leaves the better they remove the worms, and with less inconvenience are they carried.

This removal must be performed by two or three persons ; one to fill the portable trays, or two to carry them, and one who will gently remove the silk-worms from these trays upon the hurdles in the space allotted for them ; in this manner it can be executed with ease and prompt-titude.

The strips into which are arranged the silk-worms upon the hurdles, should occupy about half the space of them. I mentioned that the worms that are to occupy 174 feet of hurdles must be placed in the middle of a space of about 412 feet six inches square.

When all the silk-worms that are revived have been successively removed, there remain still some torpid upon the 174 feet of hurdle, torpid, or being just roused, that have not the strength to climb upon the shoots or branches of leaves.

The early-roused worms being removed into the great laboratory, it will be seen that they have eaten all the leaves on the young shoots and bunches of leaves that served to carry them, and that they remain without food upon the paper.

They should then be given 30 pounds of leaves chopped a little ; with these leaves, the intervals between the young shoots should be filled, and form the strips into regular order, by sweeping into their place any boughs or leaves that are scattered irregularly, with the broom made for this purpose.

After this second meal, those worms that were heaped up together will be seen stretching out evenly.

The other 30 pounds of leaves should not be

given until the second meal has been thoroughly consumed; and should the young shoots and leaves not be required, they may be given the next day.

Although it is not a general custom to chop or cut the leaves for silk-worms in this fourth age, I have, however, found it very beneficial to give it to them, coarsely cut up, not only the first day, but also the second and third. I have already mentioned, that when the silk-worms are just issuing from the cast skins, they are weak, and not very hungry; fresh leaves, slightly cut up, by exhaling a stronger smell, stimulate their hunger, and the cut edges are more easy to bite.

The late-roused silk-worms should be placed on hurdles distinct from the earliest worms.

At the end of this day, the worms begin to shew some vigour; they move quickly to the leaves, they grow perceptibly, they lose their ugly colours, become slightly white, and assume more decided animal action.

When all the silk-worms are taken out of the small laboratory, the hurdles from which they have been removed should be well cleaned.

This should be done quickly, if any of the silk-worms are to be put into the small laboratory again for the convenience of space. The litter should be rolled up in the paper, and taken out, but there is no hurry to clean the hurdles, if the worms are not immediately to be put into this small laboratory again.

In the third age, about 300 pounds of picked leaves have been put upon the hurdles, and the litter removed is of about 93 pounds weight, consequently 207 pounds of substance have been consumed by the silk-worms, and in evaporation. The excrements of the silk-worm of this age weighed about 18 pounds.

Second Day of the Fourth Age.

(Seventeenth of the Rearing of the Silk-worm.)

For this day will be wanted 165 pounds of sorted leaves, slightly cut up.

The two first meals should be the lightest, and the last the most copious.

The worms grow fast, and their skin continues to whiten.

In giving the meals, the space occupied by the worms should be widened.

Third Day of the Fourth Age.

(Eighteenth of the Rearing of the Silk-worm.)

For this day will be needed 225 pounds of sorted leaves, a little cut.

The two first meals ought to be the most plentiful, the last meal to be of about 75 pounds.

Fourth Day of the Fourth Age.

(Nineteenth of the Rearing of the Silk-worm.)

This day the distribution of the cutleaves should
be 255 pounds, the three first meals of about 75
pounds each, the fourth of 45 pounds only. The
worms still get whiter, and at this period are more
than an inch and a half long.

Fifth Day of the Fourth Age.

(Twentieth of the Rearing of the Silk-worm.)

No more than 128 pounds of picked leaves will
be needed this day, because the silk-worm's hunger diminishes much.

The first meal should be the most considerable.
A great number of the worms become torpid on
this day.

The leaves should only be distributed as they
are wanted, and only on those hurdles whose worms
are perceived not to be torpid, that they should
not be wasted uselessly. The worms are this day
an inch and three-quarters long.

Sixth Day of the Fourth Age.

(Twenty-first of the Rearing of the Silk-worm.)

Thirty-five pounds of picked leaves are enough
for this day.

It is easy to find out where and in what quantities the worms need food.

G 2

Since the preceding day, the silk-worms begin to decrease in size, as they have cleansed and cleared themselves of all nutritive substances, before they sink into their torpor.

The greenish colour of the rings of their body has disappeared, and their skin is quite wrinkled.

Seventh Day of the Fourth Age.
(Twenty-second of the Rearing of the Silk-worm.)

The silk-worms rouse in this day, and accomplish their fourth age.

In generalizing this paragraph, let us suggest the following observations.

In about seven days, the worms have accomplished their fourth moulting, and cast their skins.

They have consumed in that period, 900 pounds of sorted leaves, adding to which, the picking and refuse of the leaves of about 135 pounds, we shall have an aggregate weight of 1035 pounds of leaves drawn from the tree, dividing which into five parts, will allow 207 pounds of leaves for each ounce of eggs.

It is not here that I shall state the diminution of weight sustained by the leaves, by evaporation of moisture, from the moment they are culled, until they are weighed, and put upon the wicker hurdles.   We shall speak of it further on. (Chap. XIV.)

In the seven days of the fourth age, the worms which were before about one inch long have grown half an inch in length. In this age, their weight is augmented fourfold.

After the third moulting, 144 insects weighed one ounce : it now requires only 35 to make up the ounce.

After this moulting they are of a darker colour, greyish, with a red cast.

During this age, shavings of wood should be burnt in the corner fire-places, three or four times a day; dry straw will answer the purpose also, as this is done to renew and lighten the air of the room, without particularly heating it. Should it be necessary to heat the laboratory, that should be done either with the stove, or by burning large wood in the fire-places.

When straw or shavings are burnt, the ventilators should be opened for the circulation of the air.

If the exterior temperature be not cold, and should be without wind, the doors and window may also be opened; when the interior temperature, by these means, is lowered half a degree, the windows and doors should again be closed, leaving the ventilators open, and the temperature will rise again.

Those who have, instead of shutters, venetian blinds to the windows, or bars, should open the windows, to allow the air to enter.

G 3

Those who attend silk-worms should breathe as freely in the great laboratory as in the open air; they should feel no difference but that in the heat of the interior temperature, and the latter in temperature, not in the closeness.

Therefore, should the air appear heavy, the fire of straw, or shavings, ought to be lighted, to renew the air, which is done in a moment.

In my laboratories, the interior air is more pleasant than the exterior air, from the delightful smell of the mulberry leaves.

In proceeding as I have described, the silk-worms will breathe continually a pure and dry atmosphere, which makes them most healthy.

We shall see in Chapter XIII., that the construction of the laboratories should be such, that any unforeseen occurrence that may arise to obstruct the necessary operations can be easily remedied.

CHAPTER VII.

OF THE REARING OF THE SILK-WORMS IN THE FIRST PERIOD OF THE FIFTH AGE, OR UNTIL THE MOMENT WHEN THEY PREPARE TO RISE ON THE EDGES.

THE fifth age of the silk-worm is the longest, and most decisive. It requires the knowledge of the philosopher as well as the care of the practical operator, because the art of rearing these insects cannot, as many others do, make rapid and sure strides towards perfection, without the aid of physical science.

I do not intend here to give scientifical lectures, but to endeavour to make some truths obvious, which the intelligent cultivator may easily put into practice, to protect himself in all cases from those losses to which a man even of experience may be exposed, not knowing those truths. Wherefore, previous to resuming the description of the daily progress of the silk-worms, I will give a few practical observations.

Should the worms die in the first age, the loss is trifling, because expense is not prolonged, and that the leaves provided may be sold, or disposed of: while, on the contrary, should the worms perish in the fifth age, the loss is considerable, leaves having been consumed, and labour paid, as well as other expenses; besides seeing the hopes of all that profit vanished which had been reckoned upon.

It is then very needful to know the condition of the worms in the fifth age, to learn how to manage, so as to ensure their health and strength against the attacks of atmosphere or other evils that assail them.

As the silk-worms grow in the fifth age, they are liable to three evils, which attack them according to their strength, and to their distribution in the laboratory, and may weaken them so as to cause their speedy destruction.

These enemies are—

1st. The incredible quantity of fluid disengaged every day from the body of the insects, by transpiration and evaporation of the leaves given to them. (Chap. XIV.)

2nd. The deadly and mephitic emanations emitted every day from the insect, from their excrements, from the leaves, and the remains of the leaves *.

* It is surprising to find how large a portion of mephitical, consequently deadly, air disengages itself, particularly in the fifth age, from the silk-worms, in a laboratory spacious enough to contain the worms proceeding from an ounce of eggs.

If one ounce of the dung taken from the wicker trays, be put into a bottle that can hold one pound and a half of liquid, and hermetically corked ; in six or eight hours after, according to the temperature, the atmospherical air in the bottle will be found vitiated, and totally poisonous.

To certify which, a bird may be put into the bottle when it is first opened ; it will faint and die, if left in it many moments. Or if a lighted candle be introduced into it, the candle will go out directly.   These phenomena would not occur if the bottle contained atmospherical air alone.

From this it is evident that in the fifth age, the laboratory before mentioned contains 1200 pounds of dung, which quantity may corrupt, about every eight hours, a volume of air, equal to 16,800 *Paris* pints, or bottles, that are equal to hold two pounds liquid ; and in one day this quantity of dung would corrupt a volume of air of 50,400 Paris pints.

Having thus stated the quantity of corrupt air produced by the dung in the laboratory, it must appear evident how necessary it is to get rid of it, as soon as it disengages itself, and continually and gently to renovate the atmosphere.

3d. The damp hot nature of the atmospheric air, as well as the smothering heat of the laboratory, during the fifth age.

These evils injure the silk-worm in three ways:

1. The moist exhalations, produced by the leaves and the transpiration of the insect, accumulate in the laboratory, and tend to relax the skin of the silk-worm; this organ thus loses its elasticity, puts the animal into a state of languor, decreases its appetite, alters its secretions, and makes it liable to various diseases, and even to death. (Chap. XII.)

2. The mephitical emanations disengaged from the body of the insect, and from the leaves, render the silk-worm's breathing difficult, destroy its excitability even, produce disease and destruction.

3. The dampness and stagnation of atmospheric air, increased by the moisture of the laboratory, create a great fermentation in the dung, and, consequently, disengagement of heat, which, by destroying the elasticity of the air, renders it so deadly, as, in the course of a few hours, to destroy the silk-worms entirely *.

* There is another cause of disease and death, which the author does not mention, but which may be found described minutely in the *Cours d'Agriculture*, by M. l'Abbé Rozier, and which runs as follows:—

" Mephitical air is not the only cause of the speedy death of silk-worms. Atmospherical electricity equally contributes to their destruction, in the manner that it turns milk

To these causes of sudden disorders we often
have to add another, which proceeds from the

sour, also putrefying animal bodies, particularly fish. And
there is one fact that seems to establish this opinion.

" One year I laid some thin iron wire on the edge of
some brackets, that were close together; these wires were
prolonged down the supporters of the brackets; and,
finally, all united, on the floor of the room, perforated
the wall, and were led into a cistern of water. The other
brackets were not thus armed with electrical conductors.
The season was occasionally stormy, without, however, the
sultriness which sometimes occurs. The litter of the brackets
of the laboratory was changed as often as I had advised,
so that all the circumstances were equal; and I can, without
hesitation, affirm, that on all the brackets armed with conductors the silk-worms were invariably more active and
more healthy than on any of the others; and that those
brackets that were near the armed ones benefitted by the
conductors of the others. This explains a practice extant
among the peasants, of arming the bottom of brooding nests
with old iron, which had been deemed a prejudice, and a vulgar
error, by grave authors, who had not tried the experiments."
There is no doubt that too strong an impression of electrical
fluid upon silk-worms, in certain atmospherical variations,
may disorder them, and even kill them. It appears to me,
that the action of this strong natural agency, particularly in
and upon organic bodies, is not sufficiently observed. There
is much coincidence between it and the element that regu-
lates life. It is found in the interior of the globe, upon its
surface, in all organized beings that inhabit it, and in the
regions of air that envelope it. Numberless phenomena in
geology, natural history, and meteorology, depend on it.
Physicians have observed its connexion with the nervous
system, have considered it as a remedy, and have applied
it with success in some cases, but failure in other cases has
contributed to throw it, unjustly, into disrepute. My daily
practice has convinced me, that it should in medicine be con-

silk-worms being too closely distributed on the
hurdles, particularly in the last age.   This insect,

sidered under a new, but not less interesting, aspect.   I
have seen pains, vulgarly called rheumatic, sometimes disappear without any evacuations, in the course of an hour,
either by the application of certain remedies, or by friction,
or by plasters.   Can it be said that the pain proceeded
from too much excitation, and the remedy has allayed it ? or
that it was want of excitation, and that the remedy has
given it ? Surely not, since other sedative or exciting remedies failed in producing this effect.   This phenomenon then
depends on a distinct quality in the remedy applied.   Either
the pain arose from a superabundance of a certain impal-
pable fluid, which appears of the nature of electricity, and
the remedy, by attracting it, delivered the part affected by
the pain ; or, the pain arose from a deficiency or diminution
of this fluid in the organ affected, and the remedy may have
communicated a portion of the fluid contained within itself,
and thus have removed the pain.   The induction drawn
from this may be hypothetical, but the facts are certain ; and
there is no practitioner that may not have observed, that in
applying, for instance, the plaster of cantharides, or Spanish
flies, as a rubefacent, upon an affected part, it will make
the pain cease sometimes in less than an hour after the application ; and prescribing frictions, or embrocation with a preparation of cantharides, has produced the same phenomenon ;
as also by dry friction with the brush or flannel.   It would be
desirable to make experiments to ascertain how remedies act
when put in direct contact with the electrical fluid.   I am
persuaded, that were this subject to be well investigated, a
point so entirely new would be very beneficial in medical
science.   I have, in my treatise on small-pox and chicken-
pox, urged physicians to inquire into it.

   I believe electrical fluid to be one of the necessary elements which constitute, and bring into action, the organic
tissue ; that it is the principal agent in the production of the
phenomena which arise from the contact of medical sub-

as I have before stated, does not breathe by the mouth, but by small apertures, which are placed

stances with the animal economy in the morbid state, and that its great action depends on the augmentation or abstraction of the electrical fluid.

My friend, Dr. Bellingeri, a distinguished writer of Turin, who, from the period of his first studies, has applied himself to ascertain the connexion of the electrical fluid of the atmosphere with divers organic substances; or which arise from it, in physiological or pathological cases; has already published very interesting memoirs on the electricity of the blood in the morbid state; on that upon urine when healthy, and when affected by disorders; on solid animal, and on mineral liquids. For these last he has observed that they should be distinguished into three classes: into liquids which offer positive electricity, into those that present it as negative electricity, and, lastly, into simple conductors.

In the first class, he arranges alkalies, earths, and sulphurics; in the second class, the acids; and, in the third class, water, and the various solutions in which this liquid does not change its electric fluid, but which is always balanced with atmospheric electricity; and which may, consequently, be considered as simple conductors of atmospheric electricity. He mentioned to me, some months ago, having made some experiments upon some solid medical substances. I have long suspected that the principal action of tartar emetic, in the inflammatory state of the system, is to abstract the vital fluid, too much augmented, and accumulated. I will quote in support of my views the success with which Mr. Pelletan, medical professor at Paris, is practising acupuncturation, and that when the needle is run into the flesh it forms an electric current; and as the most acute pains disappear in an instant, it may be thought that they were caused by an accumulation of electric fluid upon the affected nerves, which the needle, as a conductor, has attracted into the common reservoir.

I hope my readers will excuse my having placed here a

near its legs, and which are called *stigmata.*
(Chap. II.) These breathing vessels are almost
all stopped up, and covered when the worms are
heaped together, which makes their breathing
difficult, and the transpiration also ceases, much
to the detriment of the insect. (Chap. XII.)

If these causes of disease are not well ascertained, and rectified as soon as they appear, we
shall find, when we were in hopes of gathering an
abundance of silk, that the whole crop is destroyed ; and of this, unfortunately, we have very
frequent example in all countries, and in all climates.

It is, therefore, very important to know accurately all the symptoms of disease in silk-worms,
and when they are discovered, to be able to treat
them ; although I can nearly engage that no
disease will appear, if the silk-worms are managed according to my directions for the fifth
age.

In this chapter I shall speak—

1. Of the hygrometer or barometer, the instrument with which the degrees of moisture of the
air may be measured in the laboratory.

2. Of the bottle for purifying the air, and for
drying the excremental matter deposited on the
wicker hurdle.

digression which is foreign to my object, and which can only
prove interesting to medical men.—*Translator.*

3. Of the manner of easily drying leaves in rainy weather.

4. Of rearing silk-worms until maturity.

1. *Of the necessity of the Barometer to measure the humidity of the Air in the Laboratory.*

We are surrounded by bodies which sometimes absorb the moisture of the atmosphere, and sometimes create the moisture; this phenomenon is continually passing before our eyes.

We see that salt, for instance, is more or less damp, according to the state of the atmosphere; that is to say, in proportion as it has absorbed the moisture, or lost it by evaporation.

Atmospherical air is generally dry when the wind sets from the north, and damp when from the south.

Scientific men have thought it useful to invent instruments, fitted to measure the quantity of moisture which the air may contain in any circumstances, using in their construction bodies which attract the dampness from the air easily, and by degrees, and which give it out again when the air is dry.

These bodies, which stretch when they receive the moisture, and shrink on losing it, placed in certain instruments, shew, by degrees, the quantity of moisture which they lose, or imbibe. These instruments are called barometers,

or hygrometers, or hygroscopes, all which terms
mean measurers, or indicators of moisture. As it
has been observed generally, that dry air is accompanied by fine weather, the hygrometer is, therefore, used in some places as a weather-glass.

I shall not go farther into the peculiarities of
these instruments, nor the materials with which
they may be constructed; I will only say that they
are very useful in the art of rearing silk-worms,
and that they are to be had anywhere. By
placing this instrument in the laboratory, it is
most easy to see when the air is too damp, and
to alter this by employing the remedies I have
suggested for expelling the heavy air, and re-
placing it by light fresh exterior air, which never
can be so damp as the interior.

It would be desirable to have two hygrometers
in the laboratory, placed within a certain distance
of each other, to ascertain the various degrees of
moisture in different parts of the laboratory.

There are hygrometers * joined with the ther-

* I cannot sufficiently urge the use of these instruments to
those who rear silk-worms, which indicates with such ease
one of the most powerful enemies of the silk-worm in the laboratory.

I trust I may not be imagined to recommend too many
utensils or instruments; I think I have mentioned only those
that are indispensable ; without these instruments it would
be impossible to distinguish in the laboratory,—

1st. That the temperature is not only lower near the apertures, and higher near the stoves and fire-places, but also that

mometer; but if these instruments should be thought too expensive even to effect so important an object as the regulation of the air for silk-worms, a plate of kitchen salt, coarsely pounded, may answer in some degree.

When the hygrometer indicates a very damp state of the atmosphere, or when the salt appears very damp, wood-shavings should be burnt, or straw (Chap. VI.), in the fire-places, to absorb the humidity, and replace it by the external air,

it is lower near the wicker hurdles which are next the pavement or floor, than it is on those above.

2d. That the temperature is less exposed to alteration in the higher than in the lower parts, which is the reason that the worms generally thrive better on the higher, than on the lower trays.

3d. That the dampness predominates in the lower parts of the laboratory.

4th. That the air is renewed with more difficulty in the angles of the laboratory, when there are no fire-places, than in any other part.

5th. That the silk-worms and the cocoons constantly succeed best in those parts of the laboratory, where there is a continual, well-regulated, and slow current of air.

6th. Finally, that without the instruments recommended, it would entirely depend on those who attend the laboratory, as I said in a former note, to conceal from the master the degree of temperature, either too high or too low, to which they may, through negligence, have exposed the laboratory.

All these advantages appear to me most valuable, and give a character of precision and regularity to the art of rearing silk-worms which had not before been known. There are many other arts founded on more simple rules, independent of accidental causes, which do not, however, offer so steady a degree of precision.

which is dried by this same blaze. I say blaze, and not fire, for two reasons: the first is, that, for instance, with two pounds of shavings, or of dry straw, there can be attracted from all points towards the chimney a large body of air, which issues at the flue of the chimney. While, in the meantime, this air is replaced by another quantity of exterior air, which spreads over the wicker hurdles, and revives the exhausted silk-worms. This change of air may take place without effecting any material variation in the degree of heat in the laboratory. If, on the contrary, thick wood were employed, it would require more time to move the interior air, ten times more fuel might be consumed, and the laboratory would be too much heated. The motion of air is, all circumstances being equal, proportionate to the quantity of blaze of the substances that burn quickly. When wood-shavings or dry straw cannot be got, small sticks of dry and light wood may answer. As soon as the flame rises, the hygrometer shews that the air has become drier, and the degrees of it can be seen distinctly.

The second motive which should lead us to prefer the blaze, is the light it diffuses. It cannot well be imagined how beneficial this light is, which penetrates everywhere, nor how much it influences the health and growth of the silk-worms. We ourselves often, when chilled by cold, or fatigued,

or exhausted, feel revived by the cheerful light of
a bright fire, whilst the mere heat of the fire,
without the blaze, never produces this effect.

We must then conclude that the fire of large
wood, or of the stove, is always useful when it is
required to maintain a steady temperature in
a laboratory, and that the air of it is not very
damp; but that when we want to renew the air,
charged with too much humidity, and replace it
quickly with exterior air, we should use the blaze.
When I particularly describe the laboratory, I
shall enlarge on this subject. (Chap. XIII.)

Hitherto I have spoken of the humidity disengaged from the air of the laboratory, of which I
shall give a calculation (Chap. VIII. § 7.); but I
have not yet mentioned the moisture with which
the external atmosphere often is loaded.

A barometer, placed in a contiguous room, or
outside, will shew the state of the atmosphere.
Should it be damp it would increase the dampness
of the laboratory, and this should be rectified by
frequently-burning blazes in the fire-place, to make
the air more pure and drier than the exterior atmosphere. Small fires should also be often lighted, not to communicate great motion to the exterior atmosphere, but only to keep up a gentle and
gradual agitation in the interior air, which is most
beneficial to the silk-worm. In keeping up a gentle
and constant motion of the air, the worms derive

as much benefit as if the air were drier. When it circulates freely, and is of a gentle temperature, it does not become so easily charged with moisture.

The thermometer will then shew whether the temperature of the laboratory may not require a fire of large wood, to maintain the degree of heat appointed for this age, which is the most important.

In our climate the deficiency of damp air never can occur. To absorb the great evaporation of fluid which constantly exhales from the silk-worms, the leaves, &c., the air should be invariably dry.

If north winds prevail during the existence of the silk-worms, and particularly in the fifth age, it is rare to see them fail, even in the hands of the most ignorant peasants, because the dry air absorbs the moisture exhaling from the worms and the leaves, and draws it off. This air penetrates thoroughly even into closed rooms, and absorbs the moisture of all substances, because it has a peculiar power of attracting water, as we may constantly experience in our houses.

I have observed that the greatest losses in silk-worms, under the care of ignorant cultivators, occur in the fifth age, by reason of the air becoming damp from the prevalence of some south wind, which is fatal to the silk-worms. The insects are thus plunged in hot vapour, which quickly weakens them, checks their transpiration, and kills them, although an hour before they may have looked well, and were nearly ready to rise to the boughs. In high regions, where the air is always drier and more active, these losses are less frequent.

I conclude this paragraph by observing, that the hygrometer gives the cultivator timely warning, whenever the laboratory is in danger, that those easy precautions may be taken which will ensure its safety, by removing the peril *.

2. *Concerning the Bottle which Purifies the Air of the Laboratory.*

The application of physical science to theoretical

---

* The Canon Bellani, of Milan, a philosopher of whom M. Dandolo speaks, page 46, has invented a useful instrument for clever cultivators, which he calls an Eudiometer. This term explains its use, as it indicates the exact degree of the purity of the oxygen gas contained in atmospherical air. By the help of this instrument, it is easy to discover if the strata of air immediately above the bed of the silk-worms contain sufficient vital air, or if they are loaded with fixed air, or carbonic acid gas. With the Eudiometer at hand, the cultivator may discover, at any moment, the vitiated state of the air in any part of the laboratory. It is not necessary, says M. Dandolo, that every cultivator should possess this instrument, but it would always be advantageous to an enlightened one, who should know that the art of rearing silk-worms is susceptible, in time, of reaching great perfection, which will never be the case without the aid of physical science, and the instruments which so greatly promote the success of the crops.

agriculture, may greatly and speedily destroy inveterate errors, and procure rapid improvement.

Hitherto, for instance, it has been reckoned a good method of purifying the interior air of a laboratory to burn some odoriferous or vegetable substance, to produce a grateful smell, while, instead of purifying or improving the air by these means, they were rendering it considerably worse. It has been erroneously imagined that what usually occurs in our perception of offensive effluvia, should be equally applicable to the noxious qualities of the air, which, as they affect the lungs, have great influence on the general system of animal life. The case is, however, dissimilar; in producing a pleasant smell in a room of which the air is vitiated, we do but disguise to the sense the bad quality of the air we breathe, but the lungs are not the less affected. We are then mistaken in employing such means in the laboratory; instead of correcting the noxious air, we make it worse. It might perhaps be useful here to exhibit some principles which belong more properly to physical science than to the art of which I treat; but I imagine it may suffice to point out some positive facts, from which science itself has deduced its principles.

1st. In whatever manner any odoriferous vegetable may be burnt in the centre of the room, and not in the grate ; and however grateful the odour may be, it will consume a part of the respirable or vital air contained in the room, and consequently must injure the air.

2d. The vegetable substance in burning will consume vital air, and in return exhales a mephitic gas, most noxious to respiration, and which may kill the silk-worms very soon.

3d. Vinegar even, when poured upon hot surfaces, is decomposed, and emits a mephitic gas, which augments that which previously existed. These evils which I have been relating may be mitigated perhaps by a concourse of three circumstances.

1st. The vegetable substances are sometimes burnt in the grate, or fire-place, instead of the centre of the room : in which case the effect will be like that produced when we make a blaze.

2d. At other times, when the vegetable substances are set on fire, the apertures of the apartments are all well set open, and the motion of the exterior and interior air expels some parts of the vitiated air, and diminishes the evil that might have injured the silk-worms, had the rooms been shut up.

3d. The surfaces upon which the vinegar is poured are sometimes only warmed, and not red-hot, which prevents the decomposition of the vinegar, and, by merely reducing it to vapour, does not render it so noxious as the elements of which it is composed.

From what I have stated, the cultivators, sometimes seeing their silk-worms after these operations more stimulated, have decided that the means employed were useful remedies, ignorant that the apparent advantages proceeded from other causes. At all events it is a clear fact, that perfumes tend to corrupt the air ; besides, when a laboratory is carefully attended, it will always offer a pleasant smell exhaled by the leaves, and need no better perfume, than that it should be well kept.

I should here speak of the harm which may be done by the smoke of chimneys, which spreads often through the laboratory, and remains stagnant in it. When this happens, it is caused either by the ill construction of the chimneys, or want of care in the laboratory. It may be, that the smoke is the effect of some impediment between the exterior and interior air, and may prove injurious to the worms ; although sometimes this impediment, by producing agitation in the interior air, may in some degree have its use, without those who have the care of the silk-worms being able to solve the question. Yet is it very certain that if the smoke often infests the apartment, it is to be feared we may see all the silk-worms of a laboratory perish in a moment, and the following is the way it may happen.

If the smoke is caused by the exterior air forcibly driving it down the chimney into the laboratory, and if this air, already mephitic, is heated in driving down the chimney by the fire, it may, if it is not quickly expelled, occasion the instant suffocation and death of the silk-worms, particularly should there hang any damp about the laboratory, which is unluckily too frequently the case.

Another cause of the corruption of the air in the laboratories, is the darkness in which they are generally kept ; the greater the obscurity the greater the exhalation of deadly air from the leaves of the mulberry, as would equally be the case from any other vegetable.

Should no other inconvenience result from keeping silk-worms exposed to the sun while feeding, they would have the advantage of being in the midst of vital air; because the leaves, which, when in darkness, exhale a deadly gas, would in the sun send forth the purest possible air, until the leaves were dry or consumed*.

To the harm caused by darkness in a laboratory as it vitiates the air, may be added that produced by the artificial lights employed in them. This

* There is in the order of nature, a certain and very surprising fact; when the leaves of vegetables are struck by the

series of causes of deterioration of the air which the worms must breathe, may be termed a continued

sun's rays, they exhale an immense quantity of vital air necessary to the life of animals, and which they consume by respiration.

These same leaves, in the shade, and in darkness, exhale an immense quantity of mephitic or fixed air, which cannot be breathed, and in which animals would perish.

This influence of the sun does not cease, even when the leaf has been recently gathered ; on the contrary, in darkness, gathered leaves will exhale a still greater quantity of mephitic air.

Place one ounce of fresh mulberry leaves in a wide-necked bottle, of the size of a Paris pint (containing two pounds of liquid), expose this bottle to the sun about an hour afterwards, according to the intensity of the sun ; reverse the bottle, introduce a lighted taper into it, the light will become brighter, whiter, and larger, which proves that the vital air contained in the bottle has increased by that which has disengaged itself from the leaves ; to demonstrate this phenomenon more clearly, a taper may be put into a similar bottle which only contains the air which has entered into it by its being uncorked.

Shortly after the first experiment, water will be found in the bottle that contained the mulberry leaves ; this water evaporating from the leaves, by means of the heat, hangs on the sides, and runs to the bottom, when cooling ; the leaves appear more or less withered and dry according to the quantity of liquid they have lost.

Put in another similar bottle an ounce of leaves, and cork it exactly like the former ; place it in obscurity, either in a box, or wrap it in clothes, in short, so as totally to exclude light ; two degrees after, according to the temperature, open the bottle, and put either a lighted taper, or a small bird into it, the candle will go out, and the bird perish, as if they had been plunged into water, which demonstrates that in darkness the

H

conspiracy against their health and life, and their resisting it, and living through it, shews them to be of great strength of constitution. Let us now mention the means of purifying the internal air of the laboratory, and of neutralizing and destroy-

leaves have exhaled mephitic air, whilst in the sun they exhaled vital air.

I do not think it can be necessary to make the exact calculation of the deterioration of the air caused by more or less vegetable substance, when exposed to darkness. The two extremes I have just proved by the experiment, are sufficient to give an idea of it.

Besides, light volatilizes any watery vapour with which it is in contact ; there is therefore no doubt, that all circumstances equal, the air of a well-lighted laboratory will be drier than that of a laboratory which is dark.

Many think that light is injurious to silk-worms. It is certain that in their native climate it does not injure them, although they are exposed to it by various circumstances ; however, there is here no question of exposing them to the sun, but only of rendering their habitations as light as our own.

I have always observed, that on the side on which the light shone directly on the hurdles, the silk-worms were more numerous and stronger than in those places where the edge of the wicker hurdle intercepted the light, and formed a shade, which is my reason for having very low edges to the wicker trays: any body may make this observation.   I have even seen the sun shining full on the worms, without their seeming annoyed by it. If the rays had been too hot, and shone too long on them, they might have suffered ; but this could not occur, nor does it affect the question: as I do not propose exposing the silk-worms to the sun, but only desire to shew that the air is more vitiated, and that there is more damp in a dark laboratory than in a light one.

ing in some degree the poison which exhales from
the fermented substances on the wicker hurdles,
and to produce the desiccation of those that are
inclined to ferment. I must first observe, that this
remedy will not cost above 30 *sous* for a laboratory of worms proceeding from five ounces of
eggs.

Take six ounces of common salt, mix it well
with three ounces of powder of black oxyde of manganese; put this mixture in a wine-bottle, with two
ounces of water, cork it well with a common cork.

Keep this bottle in any part of the laboratory
farthest from the stove or fire-places. In a phial,
put a pound and a half of sulphuric acid, vulgarly
called oil of vitriol, and put this phial near the
other bottle, with a small liqueur glass, and an
iron spoon. And this is the manner of using it.
Put into thesm all liqueur glass, two-thirds of a
spoonful of oil of vitriol, pour it into the large
bottle, and there will issue a white vapour. The
bottle should be moved about through the laboratory, holding it high up, that the vapour may be
well spread in the air.

When the vapour ceases, the bottle may be
corked, and replaced; even should there be no
perceptible difference between the interior and exterior air, during the fifth age of the worms, it is
good to repeat this fumigation, three or four times
a day, in the manner I have just explained. When

repeating the fumigation, the quantity of oil of vitriol poured into the large bottle may be diminished. The stated quantity of ingredients will be sufficient for a laboratory of five ounces of eggs. The bottle may be left open an hour or two in the last days of the fifth age of the silk-worms; and placed here and there in the laboratory, and even on the corners of the wicker hurdles, to diffuse the vapour thoroughly.

This remedy may be employed whenever on going into the laboratory the air appears to have an unpleasant effluvia, and that there is any closeness, or difficulty of breathing.

1st. It may take place when the litter of the silk-worm is removed, particularly in the fifth age.

2d. When in damp weather the air of the laboratory continues damp, even after having made the blaze, which renders the fermentation still quicker.

This fumigation may be of use also towards the end of the fourth age, if the air were perceived to be impure. I, however, never needed it, until after the fourth age of the silk-worms, and at the beginning of the fifth age. I use a fumigating apparatus much more convenient than the bottle, and which I shall describe in Chapter XIII., among the utensils required in the laboratory.

If there are several small fire-places in the laboratory, and that   bes are frequently made in

them, to agitate the air, fumigations will not be
so much required.

I must observe, care should be taken not to drop
any of the oil of vitriol either on the skin or clothes,
as it burns, and to hold the bottle above the height
of the eyes and nose, when it is open, because the
vapour is very searching, and would be dangerous
and unpleasant*.

Should the substances in the bottle harden, a
little water may be added, and stirred with a small
stick.

This easy remedy is more powerful than all perfumes commonly used, and produces five advantages in the laboratory.

* Although the fumigating bottle has been generally adopted, and that the bottle composed with the ingredients I have
mentioned is that used by my tenants, and which I myself
prepare for them ; yet it may not always be possible to procure manganese, nor to get it well pounded; I therefore propose a more easy method of producing nearly as good an effect.
Put ten ounces of nitrate of potash (common nitre of the
shops) into a bottle, instead of the common salt, used in the
former composition, and follow up the other recipe, only
using a smaller quantity of oil of vitriol; instead of ten ounces
of oil of vitriol for a laboratory of five ounces of eggs, it will
be sufficient to use only eight ounces. The nitre should be
quite damp, and a smaller quantity of vitriol may be poured
into the bottle to make the fumigation than I have indicated.
The gas proceeding from this last composition, is very similar to the other ; it is less subtle, and not so dangerous. It
is composed of vital air and nitrous vapour; it quickly
destroys all animal exhalations which may exist in the atmospheric air.

1st. The vapour, in spreading, immediately destroys any unpleasant effluvia.

2d. It diminishes the fermentation of the litter, and dries it up.

3d. It neutralizes the effect ofall the miasmàta, and deleterious emanations that might attack the health of the silk-worms.

4th. It revives the silk-worms, gently stimulating them, because it is composed in a great measure of pure vital air.

5th. This vapour is not alone favourable to the health of the silk-worms, but influences the goodness of the cocoon, as I have experienced.

If the laboratory is small, the remedy will be in smaller proportions, and therefore cost less.

3. *Of the Manner of easily Drying the Leaves, even in rainy Weather.*

Silk-worms consume such a large quantity of leaves in the fifth age, that if we do not try to overcome the difficulties in time, we might find it difficult, by reason of the atmospheric variations, to procure a sufficiency of dried leaves for their food.

Although we have not frequently heavy and continued rain in the month of June, I have however seen unceasing rain for three days at the latter end of this month, particularly in 1813, and at the moment of the chief consumption of the leaves ; and I observed that such an accident might occasion severe losses, if the leaves are not dried quickly, as I then dried those I gave my silk-worms.

In the other ages, the leaves may easily be kept two or three days: but on the days when the silk-worms are voracious, a number of persons must be continually at work, to provide for their daily consumption; and it is thus doing a great deal when we can get the leaves gathered a day or two before they are wanted, which I always do.

A number of writers say, that in cases of continued rains, nothing better can be done than to cut small branches of the mulberry trees, bring them home in carts, and hang them up in the houses, thus to dry them as well as possible.

These are errors which one writer copies from another without weighing the absurdity of the idea. In one single day of the extreme voracity, healthy silk-worms, proceeding from five ounces of eggs, will devour 97.5 pounds of leaves.(§ 4.)

In following this last system, to obtain such a quantity of leaves, it would be necessary to cut 6000 pounds weight of branches, supposing that the shoots of those trees only are to be cut which may produce leaves again in the year.

This might have been in the time when, from one ounce of eggs, only fifteen or twenty pounds of cocoons were obtained, because the worms did not hatch well, or died in their different ages ; but it can no longer be done so now, since from one ounce of eggs are drawn, at present, a hundred, or even a hundred and twenty pounds of cocoons.

The small branches may be cut when there is not much leaf required, as it happens until the accomplishment of the fourth age, or when there is but a small laboratory to serve.

Besides the inconvenience of having such leaves, those that gather them are liable to be wet through ; they should always have clothes prepared for them to change, and be given firing, food, and drink; lest in preserving the silk-worms we should lose the men, or at least expose them to disorders.

To dry in a day several hundred pounds weight of mulberry leaves, I proceed in the following manner.

When the wet leaves are brought in, I have them spread on brick floors, or on mud floors, which should be as clean as possible.

Then, according to the quantity, one or two persons spread them with wooden forks, turn them, throw them about, and move them much. This, often repeated, very soon shakes off the wet. If the floor is not of bricks, and the ground gets wet, the leaves should be raked off to another and drier part of the floor.

Although the leaf appears quite dry after this

operation, it still contains a great deal of water in its folds, and even on its surface.

Then twenty or thirty pounds of leaves should be spread upon a large, coarse sheet; and doubling it into the shape of a large sack, two persons should hold the four corners, and shake the leaves well about from one end of the sheet to the other, until they appear to be quite dry, which will be the case in a few minutes.

If the sheet be weighed before and after this process, it will be found considerably increased in weight by the water it has abstracted from the leaves.

Should it be required further to dry the leaves, burning a good heap of shavings, and some fagot sticks, and placing the leaves nearly all round the fire, taking care to turn them well with pitch-forks, they will become by these means as dry as if gathered at noon on a fine day; it may be affected as is required in either way; I have tried both methods, and have been equally satisfied with both: should the leaves be only wet with dew, drying them with the sheet will be sufficient. I must observe upon this subject,

1st. That even if the worms be forced to fast during some hours, to allow time for drying the leaves, it is better for them than running the chance of giving them wet leaves, which make their bodies exhale a greater quantity of liquid, and thus more speedily prepares the fermentation of the excremental substances.

2d. That the result of wet leaves further is, to make the interior air damper and more mephitic, and requires more care and attention.

3d. That the more strong and healthy the worms are in the fifth age, the better they will resist the pernicious effect of damp leaves, not properly dried; but should they attain the fifth age, in a weak and sickly state, the leaves would require to be of very good quality, and quite dry

4. *Of the rearing of Silk worms until the approach of their Maturity.*

It is now time to bring the worms to the period when they prepare to rise, and when they reject the food which they had lately so voraciously devoured.

First Day of the Fifth Age.
(Twenty-third of the Rearing of the Silk-worm.)

Since the preceding day, almost all the silk-worms must have accomplished their fourth moultting, or casting of skin, and be already roused.

The laboratory should have uniformly 68°, or 69½° of heat.

The silk-worms proceeding from five ounces of eggs, until the termination of their fifth age, should occupy 917 feet of wicker trays, or 183 feet 5 inches for each ounce of eggs.

The silk-worms proceeding from one ounce of eggs in the fifth age, consume about 1098 pounds of sorted picked leaves, which makes the quantity of leaves required by the five ounces to be 5,490 pounds weight.

In this first day of the fifth age, (which as has been said, commences on my system, in the afternoon,) the worms should fill a space of about 508 feet square of wicker, which added to the 413 feet which they already occupy, and which should now be cleaned, form together the 921 feet square of wicker, upon which they are gradually to spread until the termination of this state.

It is rather troublesome to change the silk-worms of the 413 feet, and to subdivide them, when placing them in the 921 feet, but this operation may be perfectly executed in four hours by seven or eight persons.

In the first day, about ninety pounds of young shoots, or of common leaves not sorted, will be required, and as much picked sorted leaves.

The young shoots should be directly distributed upon five or six wicker trays; and should the shoots fail, bunches of leaves may be substituted, as I before directed.

As soon as the shoots are loaded with worms, they should be taken off, and put upon the little portable trays. If the silk-worms of one wicker tray are almost all roused, they will be sufficient to fill the space of rather more than two wicker trays; and there should be formed a space in the middle of the two wicker trays, of about half the width of the tray.

When 508 square feet are filled, the trays that are left empty should be cleaned.

If in cleaning any worms should be found roused, by putting some shoots or leaves, they may be taken off like the others; should some rouse after this, they may be taken up with the hand and put with the others; but if any remain still in torpor, they must be cast away.

The sheets of paper with the litter must be rolled up, as was done in the former age, and poured into the basket prepared for this (Fig. 24), which is carried out at once.

In observing the litter when it is removed into a dry place, should some roused worms be found in it, they may be placed distinct from the others, in the warmest parts of the laboratory, with more space, that they may thrive faster, and be even with the early worms.

The litter removed will appear constantly green, and be without any unpleasant smell; but notwithstanding this, while the wickers are cleaning, the fumigating bottle should be passed two or three times through the laboratory.

It must be managed, that the worms should cover rather more than half the space which is al-

lotted to them and; thus does this operation end,
when all the wickers are occupied, having a large
space down the middle of them. In general, the
quantity of young shoots and leaves that I have
fixed will be found sufficient for the removal of
the silk-worms; but this must be regulated by
practice, and more taken if needed.

Of the six persons, which is the least number that
can execute this operation, one or two of the most
handy and neat should be directed to lift and put
the silk-worms on the portable trays; two should
carry them, and one should remove and place them
on the wicker, while the others roll the papers
and litter, clean the hurdles, and carry out the
dirt.

If it is judged necessary, another person may
be employed in distributing shoots to the later
silk-worms, that are but just rousing, that all
things may proceed without bustle or confusion.

Should it be deemed advisable to divide the
operation of cleaning and removing, it may be
done by cleaning only half the wickers in the
morning, and the other half in the evening; in
which case, the worms, whose changing is de-
ferred to the evening, must have one or two
meals given them. Although this manner of
proceeding is not bad, yet I prefer changing
them all at once, and it may be done in four
hours, when the worm is steady.

The ninety pounds of shoots and leaves on

which the silk-worms were removed, furnishes them with an abundant meal; the other ninety pounds of sorted leaves should be divided into two meals, which should be given them every six hours. In giving the first meal, care must be taken to straighten the lines of the strips on the hurdles, by sweeping any straggling leaves or worms into regular order with the little broom.

At the third meal, the strips should be widened a little. Should there be too many worms in some parts, they should be taken to cover the barer parts of the wickers.

The silk-worms appear tolerably strong this first day.

At the fourth age, 900 pounds weight of leaves were distributed on the wickers, and the litter of this age weighed 300 pounds. The silk-worms have thus derived 600 pounds weight of the substance, including the loss by evaporation. The excrement weighs about 93 pounds.

If the exterior temperature be mild, and little different from that of the laboratory, it might be left open while cleaning, and thus admit on all sides a free current of air; and also burn shavings to make a blaze, and this is particularly necessary, when the cold or dampness of the weather precludes opening all the apertures of the laboratory while cleaning. In cases of cold and high winds, the upper and lower ventilators may be kept open, which will renew the air as

much as the blaze. And in all cases the thermometer and hygrometer must positively regulate all proceeding by their indications.

Second Day of the Fifth Age.

(Twenty-fourth of the Rearing of the Silk-worm.)

For this day will be wanted 270 pounds of
leaves sorted, divided into four feeds ; the first,
which should be the least of about 52 pounds,
and the last, which is the most plentiful, of 97
pounds weight.

In distributing the food, the strips should be
widened gradually.

At the close of this day the worms are much
whiter, and considerably developed.

Third Day of the Fifth Age.

(Twenty-fifth of the Rearing of the Silk-worm.)

This day the silk-worms will require about
420 pounds of sorted leaves. The first feed
should be of 77 pounds ; the last feed should be
the largest, and of about 120 pounds weight.

The worms continue to whiten, and many
appear upwards of two inches long.

They could eat on this day a larger quantity
than I have specified ; but I think it most
beneficial not to add to this quantity, that they
may thoroughly digest it ; besides which, it

strengthens their constitution, and makes them livelier. The strips they occupy should be widened whenever they are fed.

Fourth Day of the Fifth Age.

(Twenty-sixth of the Rearing of the Silk-worm.)

This day the silk-worms will want 540 pounds weight of sorted leaves; the first feed should be of 120 pounds weight, and the last of 150.

The worms now are beginning to grow voracious; they grow handsomer and stronger; some are two inches and a half long.

Fifth Day of the Fifth Age.

(Twenty-seventh of the Rearing of the Silk-worm.)

The worms will this day want 810 pounds of picked leaves; the first feed of 150 pounds, and the last meal of 210 pounds weight.

If necessary, the silk-worms should have some intermediate food; when the regular distribution of leaves is devoured in less than an hour and a half, the worms must not be suffered to fast five hours, but receive some leaves in the interim, particularly if there should have been wickers on which the worms had not been as well fed as the others at first. For although I have fixed the quantity of food for this day, it is always necessary to be regulated by experience: should the worms want more food, they must be given it.

In the course of the fifth age the wicker trays should be cleaned. If the litter is dry and fresh, they need not be shifted till the evening of this day, or the beginning of the second day ; but this must depend on circumstances, and the convenience of the cultivator.

Care must be taken in distributing the last meal on this day, only to feed four wicker trays at a time, to allow of time insensibly to lift off the silk-worms before they have finished eating the leaves given them.

As this time the worms are not to be removed ; the wickers must be cleaned after another manner.

The following is the manner of cleaning them. The portable trays are put on the edges of the wickers, and when the leaves are loaded with silk-worms, they are put in single layers on the portable trays. When several of these are filled, the litter, with or without the sheet of paper, must be carried off in the square baskets, (Fig. 19) which are hung near the wicker trays: the litter being removed, the paper should be swept and cleaned with light brooms ; the sheets of paper are laid down again, one after another, and the leaves with the worms replaced on them. This is repeated until the litter has been entirely changed throughout the laboratory.

When one basket is full it is carried out, and

another substituted.   Great care should be taken
not to hurt or bruise the worms in removing
them.   Six persons, at least, should be employed
to perform this cleaning of the litter expedi-
tiously, and in their number I do not include
those who carry the litter out of the laboratory.

This litter has no unpleasant effluvia, and is as
green as the leaf itself; and the paper upon
which it lay is only a little damp.

This change of the litter will employ eight
hours; and, therefore, in this period the silk-
worms that have been cleaned should be fed, and
those that are to be cleaned last, may be fed
before they are cleaned, that none of them may
fast too long.

It must not be forgotten, that during this
operation, as the case may require it, there should
be light blazing fires burnt, and the fumigating
bottle should be passed twice round the laboratory; the windows and ventilators should be
opened, according to the state of the exterior
atmosphere, but in all cases the ventilators in the
ceiling and floor, and all the doors, must be open.

If the exterior air be very damp, which would
indicate that of the laboratory being still more so,
the small blazing fires in the angle fire-places
may be frequently repeated.   And if this should
raise the temperature too much, it may easily be
lowered, by opening the ventilators and windows,

being guided by the thermometer and the hygrometer.

At the conclusion of the fifth age, we shall state the total weight of the litter which it will have produced.

Sixth Day of the Fifth Age.

(Twenty-eighth of the Rearing of the Silk-worm.)

The silk-worms should have 975 pounds of picked leaves, divided into five feeds, the last of which should be the most plentiful. The silk-worms now eat most voraciously, and some even attack the mulberries which are among the leaves.

If, after having distributed the leaves, the quantity appears insufficient upon some wickers, and thus it has been devoured in an hour, an intermediate meal should be added when necessary.

Knowing the quantity of leaves to be given in the day, it is easy to distribute them either into four or five meals, as it may appear to suit the silk-worms best. If the wickers could not be all cleaned on the preceding day, the operation may be finished this day.

The shelly black shining proboscis placed at the extremity of the muzzle is become stronger ; it is in this proboscis that are placed the small saws which tear and separate the hard leaves, and even the fibrous parts of the leaf.

Some of the worms are now three inches long; they are become whiter; to the touch they present a soft velvety surface, and are strong and healthy.

By giving more food to the worms last removed from the hurdles, and by allowing them more space, they will soon equal the earliest in size.

Seventh Day of the Fifth Age.

(Twenty-ninth of the Rearing of the Silk-worm.)

The silk-worms will require this day 900 pounds weight of well-sorted leaves. The first meal should be the largest, and those following should diminish; should there be any intermediate meals wanted, they must be given as before.

Some worms will now be seen upwards of three inches long.

The extremity of the insect begins to grow shining and yellowish, which shews they are approaching to maturity.

Some of them begin to eat with less voracity.

They this day attain their largest size, and their greatest weight.

On an average, six silk-worms now weigh an ounce.

Thus their weight has increased fivefold in seven

days since the fourth moulting; at which time thirty-three silk-worms made an ounce.

They are also grown an inch and half longer, or nearly double the length, in the same time; as, on the twentieth day, they were only an inch and a half long.

Towards the close of this day they begin to lose size and weight; because, reckoning from this day, they take less food in proportion to the quantity of excrement and steamy vapour which their bodies discharge.

We shall continue to observe them as they decrease, as we have done while they increased.

They are at present in their highest condition.

Eighth Day of the Fifth Age.

(Thirtieth of the Rearing of the Silk-worm.)

The silk-worms this day must have 660 pounds of well-sorted leaves. The proportion of leaves must diminish, as the appetite of the worms decreases much.

The food must, as usual, be divided into four, giving them the largest meals first, and gradually diminishing. The first meal of 210 pounds of leaves.

That the maturity of the worms may be perfectly alike, some intermediate food should be given, according to necessity, to bring on those worms that are backward.

During the last days of the rearing of the worms, they should be fed with the best sort of leaves, always culled from the oldest trees.

The silk-worms now advance towards maturity, which may be perceived by their yellow colour, which increases from ring to ring.

The backs of the silk-worms begin to shine, and the rings lose the dark-green colour that marked them.

The advance to maturity is also denoted, in some of them, by the diminution of their bulk in the course of this day, and by their seeking to fix themselves to the edge of the hurdles, to void the substances with which they are loaded.

This day, and more or less speedily, according as the signs of maturity increase, and that the litter gets moist, the wickers should be cleaned in the manner we before described, being very careful to take the worms gently with the leaves upon which they lie, that they may not be bruised.

Light fires, and the fumigating bottle to purify the air, the ventilators, and the use of the thermometer and hygrometer, are in this change of litter more essential than on any former change.

The smell of my laboratory is always fresh and pleasant ; while the hurdles are cleaning, the litter is always sweet and green, and a little damp.

Ninth Day of the Fifth Age.

(Thirty-first of the Rearing of the Silk-worm.)

The silk-worms this day need 495 pounds of leaves, which must be distributed as it may be wanted.

The yellow hue of the silk-worms grows deeper, their backs shine more, and in some of them the rings assume a golden appearance.

The muzzle is become of a brighter red than it was in the beginning.

From time to time a gentle fire should be lighted, particularly in the night; twice a day the fumigating bottle should be passed through the laboratory; the ventilators should never be shut when the fire is lighted, nor indeed at all, that the air may be renewed entirely.

In a well-constructed laboratory, atmospheric variations need not be feared, which otherwise might be fatal to the silk-worms. (Chap. XIII.)

Since I first reared silk-worms, I have seen them exposed to every variation of seasons, and to many accidents that might have proved injurious to them; however, my practice of rearing them was such, that I have always preserved them full of health and vigour.

Let us now take the general result of what we have stated in this paragraph. In the course of about thirty days, in which the silk-worms have attained their greatest development and heaviest weight, I have observed the following facts:—

1st. That in their growth they are become forty times larger than they were when first hatched, being then about the size of one line.

2d. That, in thirty days, their weight is become *nine thousand times* greater, since it required 54,525 young hatched silk-worms to form an ounce (Chap. V. § 3.), whereas six silk-worms, when full grown, are now sufficient to weigh an ounce.

3d. That the fifth age alone, which is their longest and most happy period, comprises two-thirds of their lives.

From the ninth day of the fifth age, and the thirty-first of the life of the silk-worm, until the completion of its maturity, we shall see, that although they need little food, they still require great care. Of which we shall speak in the following Chapter.

Reckoning the 240 pounds of sorted leaves, which are to be given on the morrow (or tenth day), the worms proceeding from five ounces of eggs will have consumed, in the fifth age, 5490 pounds of sorted and picked leaves.

Adding to this 510 pounds of refuse and pickings of the leaves, the total weight taken from the tree will be 6000 pounds of leaves.

The total weight of dung drawn from the wicker trays in the fifth age is about 3300 pounds weight, which demonstrates that from 2190 pounds, a part served to nourish the silk-worms, and the rest exhaled in vapour.

On comparing the dung produced in this age with the quantity of leaves given to the silk-worms, it will be found to be in greater proportion than that produced in the other ages. We shall explain the reason of this. (Chap. VIII. § 4.)

Calculating the weight of the leaves, and the loss by evaporation of the moisture, of which we shall speak presently (Chap. XIV.), the worms must have consumed, in their fifth age alone, 1200 pounds of leaves per ounce of eggs.

We shall see, in the following Chapter, that the worms accomplish the fifth age, and cast their skin in their transition to the chrysalis, when they have lost more than half their weight.

We then see those marks on the backs of the silk-worms which I have already mentioned ; and the shelly proboscis, black and shining, which is attached to the muzzle, acquires a very considerable force. At this period of progress the worm is whiter than ever.

As I must suppose these insects to have been reared upon the system I have described, they will have been uninterruptedly sleek and healthy.

I have already said, that should it be impossible, from the heat of the season, to keep the temperature at the degree fixed, it must be as nearly kept as may be possible, by using all means of cooling the air, but *never* excluding it from free circulation in the laboratory.

If, in airing the laboratory, too great a body of cold air should be allowed to enter, it will harden the silk-worms a little; and there is nothing to be done, but when the air is changed, to light the stove and grate fire, and again bring up the temperature to 69°, leaving the ventilators open. The skin of the silk-worms will soon resume its softness, and this check will have done them no injury.

The silk-worms, to the touch, always appear cooler than the atmosphere of the laboratory; their transpiration is the cause of this. Although it is insensible, it is very abundant; for as they void no liquid except when they rise, transpiration is the only means they have of evacuating the moisture they swallow with the leaves.

Our own skin is never cooler than when it is in a state of moisture, exposed to the impression of the air; evaporation and cold augment in proportion to the action of the air, although the temperature may be rather warm.

It would be dangerous not to keep the air of the laboratory in motion; because it could not otherwise carry off and absorb the quantity of

water and vapour that exhales from the bodies of
the silk-worms, besides that moisture which proceeds from the leaves, excrements, &c. &c.

Chapter VIII.

OF THE REARING OF THE SILK-WORM IN THE
LAST PERIOD OF THE FIFTH AGE; THAT IS
TO SAY, UNTIL THE COCOON IS PERFECTED.

Observations on the Subject.

LET us for a while leave the silk-worms on the
wickers, to mention some things relative to them,
and to shew the preparations we should make for
the accomplishment of their fifth age.

The fifth age can only be looked on as termi-
nated when the cocoon is perfect. When the
silk-worm has poured out all its silk, and formed
its cocoon, it casts its envelope, and becomes a
chrysalis. (Chap. I.) But to form the cocoon, it
must attain to this point, viz. of becoming a compound of only two remaining substances, the silky
substance and animal substance.* It must then

---

* Besides these animal and silky substances, which almost
wholly compose the silk-worms, there are in their organs
earthy particles, alkaline and acid substances, part of which
are even in solution, as I shall hereafter shew. These substances do not act upon each other when they are in small

have unloaded itself of all the excremental matter contained in the intestinal tube.

It is not only requisite to know the last degree of perfection of the worm, to facilitate its means of forming the cocoon, but also to know all the other operations necessary to ensure the cocoons being of a very good quality.

The cleanliness of the tables, in these last days of the fifth age, requires great attention, to preserve the health of the silk-worms.

It is with these insects as with all other animals, some are quick in all their operations, others more slow ; and it is important to form just observations on these facts.

To enable those who rear silk-worms to have correct ideas on the necessity of maintaining dry and pure air in the laboratory, I will endeavour here to convince them, by evident facts and calculations, that although it may appear to them that moist vapours and mephitic exhalations have ceased in the laboratory, it is, on the contrary, at this moment, that they prevail in the

proportion ; and when by vital action, or the disposition of the organs, they are balanced and suspended, the chemical law of affinity cannot be applied to them. But when these substances accumulate for want of care, and vital action diminishes, then, as we may see in Notes 20 and 21, these substances will produce re-action in the body of the silk-worm, destroy the equilibrium which existed, and produce those diseases which are the evident result of chemical action and attraction.

greatest abundance; an incredible quantity exhaling from the body of the silk-worm while it is making its cocoon, and even after the formation of the cocoon.

I must add, that if the quantity of leaves I appointed for the tenth and last day of the fifth age be insufficient, a very little more should be allowed them; for they should now be stinted, even if there chance to be some leaves left. And also should the worms take eleven days, instead of ten, to come to perfection, the same quantity might suffice. There are causes we cannot trace, which hasten or slacken the progress of the silk-worms towards maturity, by some hours.

These are the subjects of the following paragraph:

1st. Matured perfection of the silk-worm.

2nd. First preparation for forming the espalier, or hedge, for the rising of the silk-worm.

3rd. The last feed given to the silk-worms.

4th. Cleaning of the wicker tray, and conclusion of the wood-work and espalier.

5th. Separation of the silk-worms that will not rise on the espalier; and last cleaning of the wickers.

6th. Care of the laboratory until the worms have accomplished the fifth age.

7th. Quantity of excremental substance, vapour, and gas, produced by the silk-worms, from the moment they have attained their full growth, until their perfection, and until the formation of the cocoon.

1. *Matured Perfection of the Silk-worm.   Tenth Day of the Fifth Age.*

(Thirty-second of the Rearing of the Silk-worm.)

We have seen, in the foregoing chapter, how the worms begin and continue to shew signs of maturity.

This last day they attain perfection, which may be ascertained by the following indications :

1st. When, on putting some leaves on the wickers, the insects get upon the leaves without eating them, and rear their heads, as if in search of something else.

2nd. When, on looking at them horizontally, the light shines through them, and they appear of a whitish-yellow transparent colour.

3rd. When numbers of the worms, which were fastened to the inside of the edges of the wickers, and straightened, now get upon the edges, and move slowly along, instinct urging them to seek change of place.

4th. When numbers of worms leave the centre of the wickers, and try to reach the edges, and crawl up upon them.

5th. When their rings draw in, and their greenish colour changes to a deep golden hue.

6th. When their skins become wrinkled about the neck, and their bodies have more softness to the touch than heretofore, and feel like soft dough.

7th. When, in taking a silk-worm in the hand and looking through it, the whole body has assumed the transparency of a ripe yellow plum. When these signs appear in any of the insects, every thing should be prepared for their rising, that those worms that are ready to rise may not lose their strength and silk, in seeking for the support they require.

2. *First Preparation for forming the Hedge, or Espalier.*

To avoid the loss that might accrue from delay, there should be fagots or bundles, ready made, of dry turnip-plants, or common broom, or clean bean-stalks, or, in short, of any bush, or brush-wood well cleaned, that may answer.

These should be arranged into bunches or fagots, that the worms may easily climb up them, and fix themselves conveniently to pour out their first downy silk, and then work their cocoon. These bushes should be neither too thick set, nor too bare, to avoid inconveniences, which I shall mention hereafter. As soon as it is observed that the worms want to rise, the fagots should be put up against the inside wall, above the wicker

trays, on the most convenient side, leaving fifteen inches between each bundle or fagot.

The twigs, or top branches of the bundles,
should touch the lower part of the tray above
that on which it is placed, and by being bent
down by the tray above, form a species of arch.
Upon which we must observe,

1st. That the fagots should be placed a little
aslant, so that the worms that climb up may run
no chance of dropping off.

2nd. That the fagots should always be longer
than the height between the floor and the wicker
hurdles, or than the height between the lower
wicker and that above ; thus they may always
form a curve when placed between them, and in
this manner the worms that rise upon the curving
part of the fagot do not soil the worms that
are climbing perpendicularly under them, when
they evacuate, which would be the case were
the fagots not arched.

3rd. That the branches of the fagots should
be spread out like fans, that the air may penetrate through all parts, and the worms may work
with ease. When the worms are too near each
other they do not work so well, and form double
cocoons, which are only worth half a single round
cocoon. This inattention, which is almost uni-
versal, causes great loss every year, which is
little known, except by the manufacturers who

spin the silk, and are obliged to separate the double cocoons from the single, the silk being of an inferior quality.

The little fagots should be fixed into the wicker-work of the hurdles, and not into the paper, which is very easy, and requires only to lift the paper at the edge of the wicker, to put in the ends of the fagots through the wicker, so as to let the fagot touch the edges. This arrangement of the bundles or fagots, is also convenient for the cleaning of the hurdles, which must soon occur.

Having thus placed upon each hurdle, and in their angles, a sufficient number of spreading fagots, the first worms that are ready easily find their way up. If in the course of this day (which requires the very utmost care,) in watching the hurdles, some worms should be perceived ready to rise, they must be taken up and put near the ends of the fagots: this operation is useful. There should be also some dry twigs of oak, or other wood, put upon the wickers, and when the worms rise on them, they may be lifted and put close to the fagots, which will save the trouble of constantly looking for the worms that are ready to rise.

I must observe, however, on this subject, that during the first three or four hours in which the silk-worms give signs of rising, it is not necessary

i 5

to be in a hurry to make them climb upon the fagots; for by remaining some hours on the hurdles, they have time to cleanse themselves by evacuation upon the litter.

Whatever may be the method followed in the course of this period, it must always be desirable that the little bundles should be well placed, well arched, clean and light, and not thick ; that, as I before said, the air may circulate freely, and that the worms may work with ease in them.

3. *Last Feed to be given to the Silk-worms.*

The 240 pounds of sorted leaves, which are still in reserve, should be given them by degrees, and according to their wants.

The little appetite of the silk-worms, and their wish to rise upon the leaves, proves that even were they given more food at one time, it would only add to the litter, which would become dirty, because this is the period at which they evacuate most. From this, it is better rather to stint them in each distribution.

The hours of feeding cannot be fixed in this last day; it cannot even be known whether there may not be required a small quantity of leaves for the following day.

It is very obvious that latterly the digestive faculties of the silk-worms decrease, and that it is only habit or intemperance which inclines them to feed ; and as they approach the period of rising, their excrements are nearly of the taste and colour of the natural leaf, which proves that it has hardly been decomposed in their bodies.

It also happens that many worms, too full of food, afterwards require a day or more to evacuate it, and shew signs of uneasiness before they become empty, evidently proving that the functions of the stomach and intestines are consider ably weakened.

4. *Cleaning of the Hurdles ; end of the Preparations for the rising of the Silk-worm.*

As soon as the worms are prepared to rise, the hurdles should be cleaned thoroughly. This operation, although tedious, is easy enough, with the aid of the portable trays. These portable trays cannot now be put on the hurdles, because the fagots placed round them prevent it ; however, they may be supported against the trays, so as to be able to use them within. When they are placed near the trays, the silk-worms must be carefully put upon them. Two or three portable trays should be filled. This done, the litter should be emptied from the sheets of paper into the square baskets.

When one portion of the hurdles have been cleaned, the paper is replaced, and the worms gently slided down upon it by slanting the tray.

They should, strictly, only be given the quantity
of food they may want, and that very sparingly.
When the square baskets are filled with the litter,
they must be directly carried out of the labora•
tory. In this manner, several persons may clean
the hurdles in a few hours. The silk-worms,
when put on the portable trays, should be handled with the greatest gentleness and ease, leaving
them on the twigs or bits of leaves to which they
are fastened, not to hurt them in tearing them
off. The slightest injury, at this age, is particularly hurtful to them, because the vital action is
much diminished.

In sliding the silk-worms upon the hurdles,
they should be placed in squares of about two
feet, beginning on the side upon which the espalier or hedge is already placed, and forming the
squares close to them, so that the silk-worm may
find no difficulty in rising upon the fagots; a
distance of eight or ten inches must be left between the squares. In the centre of these squares
should be fixed bunches of heath, or small dry
boughs. This operation may be performed by
eight persons in eight hours.

In this cleaning of the laboratory, it may be
seen that there is a larger quantity of excrement
and litter than in the former ages, because the
silk-worms evacuate more in the last days, and
the leaves given them latterly are more loaded

with mulberries, and consist of coarse and woody substance, which they cannot eat.

During the time of this operation, the exterior air should be freely admitted on all sides, and may be drawn in by lighting a blazing light fire in the grate.

All the ventilators should be open, as well as the doors and windows; if there is no wind, and if the weather is not much below the 68th degree of temperature, which is the prescribed heat of the laboratory. Although generally the air, at this time of the year, is neither cold nor windy enough to be obliged to shut up the laboratory, it has happened to me to be obliged to take great precaution in admitting air. In June, 1813, the exterior air was at 53°. The unfavourableness of the weather lasted long, and rain and wind were continual; and I was obliged to be careful, even in opening the ventilators.

In such cases, a part only of the ventilators should be opened at once. Light fires should be lighted in the fire-places, and obtain thus a free circulation of air, gentle and steady, without cold, and getting rid of a good deal of damp, which improves the silk-worms, and enables them to breathe more freely. The fumigating bottle should also be passed once or twice through the laboratory, and the hygrometer will shew whether the air is grown sufficiently dry.

During this time, the worms continue to rise and climb, and it is thus indispensable to finish the hedge, and to fill the hurdles with rows of fagots. We stated that the first row of inside fagots should be placed at six or eight inches' distance from one another, to form the hedge ; other small fagots must be stuck in between them, and form a species of vaulted roof under the higher hurdle. It should not be too thick ; the small fagots may be stuck into the lower hurdle without taking off the paper. Across the middle of the hurdle, and between the squares into which the silk-worms have been laid, should be stuck four fagots in a bunch, well fanned and spread out to admit the air, and that the silk-worms should be able to rise and climb into every part of them to make their cocoon. When the hedge is formed round three sides of the wicker hurdles, and the groups or bunches of fagots are placed in the centre of them, the worms should with great care be put nearer the hedge, that they may climb with ease. The clumps or bunches of fagots should be about two feet from one another, and will hold a great quantity of silk-worms.

As soon as the hedge and bunches are nearly laden with worms, other small fagots should be put between the hedge and bunches, and between those bunches and the outside edge of the wicker trays. Thus are formed parallel hedges across the wicker trays at two feet distance ; and, as all the top branches wave and bend under the wicker trays above, or the ceiling, the whole presents an appearance of small avenues covered in at top, and shut in at the end of the hedge, and with us these are called *cabanes* or huts.

This arrangement of fagots will generally suffice to receive all the silk-worms of a wicker hurdle ; should there, however, remain some silk-worms on the tray, when the fagots are nearly laden, a small branch may be put against the fagots, and thus prevent their lying too thick together on the hedges. If care has been taken to provide long sweeping fagots, well curved at the top, and well spread out, that the air may pass through the fagots, the number I have described will be found quite sufficient to answer all purposes, and the silk-worms will with ease work well, not huddled together, and will not touch each other, and not produce double instead of single cocoons, which are not so valuable.

Two essential things should always be attended to. The first is, to put those worms near the fagots which are perceived to be ready to rise ; and the second is, to give a few leaves to those worms that are still inclined to eat. One or two careful persons should be thus occupied.

As long as the worms feel a wish to eat, were

it only one mouthful, they will not think of their cocoon, and it will happen, that after climbing, and even evacuating themselves, they sometimes go down again for more food. I have seen them stop when descending, and remain with their heads downward, the wish to eat having ceased before they reached the bottom; they should then be turned, so that their heads may be put upwards, as the position is injurious to them.

These attentions, which appear too frivolous, often contribute, however, to an abundant crop of the best cocoons, with few double ones.

I have often seen, in visiting laboratories that had no fault but that of ill-arranged hedges and clumps, namely, that the fagots were too thick, irregular, the air not circulating, the worms were straitened, many cocoons double, others imperfect, soiled, and some of the silk-worms smothered before the completion of their metamorphosis. In which cases, far from a fragrant smell, the air of the laboratory had a stench most offensive, produced by the decomposition of the silk-worms.

5. *Separation of the Silk-worms which will not rise. Cleaning the Wicker Hurdles for the last Time.*

Four-and-twenty or thirty hours after the worms have first begun to rise, and when four-fifths have risen, there remain on the wickers those that are weak and lazy, who do not eat, do not seem of the disposition of those that have risen, but remain motionless on the leaves, without giving any sign of rising.

The care of these silk-worms being quite different from that required by the others, they should be taken away, and put either in the small laboratory, or in any dry clean room, of at least 73° of heat, where there are hurdles covered with dry clean paper, and the hedge ready prepared for them. (§ 4.)

As soon as they are thus placed, some will rise directly, others will eat, and then rise, and so on, till all will have risen. These worms will have acquired the vigour and stimulus they wanted, by being put in a warmer and much drier apartment.

The great mass of silk-worms in the large laboratory, in evacuating themselves, often soil one another with excremental matter. Moisture, as I have frequently observed, is very pernicious to these insects, and, if once wetted, it checks their transpiration; that alone will destroy their vigour, and indispose them to rise. As fast as the hedges and clumps are formed, the worms that rise on them shed liquid matter upon the litter and paper, where lay the later silk-worms, which increases their languor and listlessness, even supposing the air to be pure ; if it were damp, they would be slower and lazier still, so the best remedy is to remove them at once to a very dry and tolerably warm spot.

Should these worms be very numerous, not only should there be the hedge round the hurdles, but also the clump and hedges across, that they may have every facility for rising offered to them.

If only a part of these worms appear inclined to rise, they should be covered with some leaves, and some twigs put over them, that when they climb upon these, they may be taken in the hand, and put upon the fagots, as they are then ready to rise.

With this assistance, the lazy worms will distribute themselves in the branches, evacuate, and begin weaving the cocoon.

Before these few worms are put on the fagot, we may form a sort of support or couch of turnip-straw for them among the branches, to prevent their dropping off, and to give them time to fasten themselves to the branches. In this manner, I have obtained cocoons from almost every silk-worm.

All the silk-worms being off the hurdles, having either risen, or been carried away, no time should be lost in cleaning the hurdles, which will be the last time of performing this operation, which must be done with the greatest expedition.   First, taking away with the hands the litter that may be lying near the edges and clumps, then removing all the excrements that may remain either with the broom or dung-shovel, (fig. 25), or any convenient tool, carefully cleaning every part of the wicker.

It is beneficial to carry off, as quickly as possible, every thing that tends to corrupt the air, or make it damp.

6. *Care of the Laboratory until the Silk-worm has completed its Fifth Age.*

1st. When the worms manifest a desire to rise, infinite care should be taken to prevent the temperature of the laboratory from falling; it should be maintained between 68° and 71° that if the exterior air be cold, it may never directly strike upon the worms, that no cold blasts should reach them; but, on the contrary, that a gentle circulation should be made from the top to the bottom of the laboratory, by means of the ventilators in the ceiling and floor, which must be opened more or less, according to circumstances; and the air may be circulated from the contiguous apartments by opening the doors into them. These precautions are needless, when the exterior air is warm, and is not agitated by much wind.

It is proved that any violent agitation of air cramps the worms, stuns them, causes them to drop off, and to suspend the work they had begun *.

* The common idea is, that any violent commotion in the air, either by the noise of thunder, or fire-arms, makes the worms drop off from the fagots. So the peasants much dread the effects of thunder, and if the silk-worms fail to rise, and thunder has been heard, they look upon it as the sole cause of the failure and loss. For the same reason they avoid making a noise, for fear of disturbing the silk-worms at their work.

But if experience be consulted (says the author of the ' Cours d'Agriculture,' M. Rozier, in the article treating of silk-worms,) it will convince us that neither the sound of thunder, nor that of loud musketry, would make the silk-worms drop off, but they will continue to work as if in the stillest solitude. The following fact will confirm this. About thirty-five or forty years ago, at M. Thomé's, a great cultivator of silk-worms, and one of the first who had written on the culture of mulberry trees and the rearing of silk-worms, we, in presence of several witnesses, fired several pistol-shots in the midst of the laboratory itself, when the silk-worms were rising and high at work; only one dropped off, and it was evident that it was a sick one that could not have formed a cocoon at any rate. And nobody can doubt the testimony of M. Sauvage, who tried the same experiment in his laboratory, without producing any effect on the silk-worms. The general idea is, then, contrary to experience, and demonstrated to be ill-founded by undeniable facts. The motion of the air, then, occasioned by the noise of thunder, is not injurious to the silk-worms that are spinning the cocoons, but the lightning and sound indicate an accumulation of electricity in the atmosphere, which discharges itself from a cloud overladen with it, upon another cloud which has a less quantity of electricity, or none; or in short, between the sky and the earth, until the electricity has found its equilibrium in the total mass. This equilibrium, however, cannot be regained without these slight insects

2d. When the worms are near rising, the air should be kept as dry as possible, that the paper on the wicker may dry, which is wet with the moisture of the excrements; and that the vapour which exhales from the body of the insect may be absorbed and carried off, the quantity of which is very considerable, as we shall prove.

3d. Should any of the worms drop off that had risen, they should be taken up and carried into the little laboratory, where the other later worms were put; to avoid that the late worms

being affected by it. Do we not see persons with delicate nerves, or overcharged with electricity, in convulsions, and even with fever, on such occasions ? Is it, then, surprising that worms filled with silk, which, it is well known, becomes electric by mere friction, although it has not the property of communicating its electricity to the objects surrounding it, —is it surprising that these insects should be much oppressed and tormented by their own electricity, and by the superaddition of that which they receive from the atmosphere ? If to this first cause is added any other, it may easily be perceived what occasions the dropping off of the worms, and this will no longer be attributed to the sound of the thunder in the air, &c. &c. Before the storm commences, the weather is lowering, heavy, and loaded, the heat so suffocating that we can hardly breathe ; the vapour seems to oppress all nature ; no breeze is felt, no leaf stirs. The animal substances putrify quickly ; in short, the sultriness is perceptible everywhere, in the atmosphere as well as in the air of the laboratory. May not the silk-worms be faint, suffocated, in these oppressive moments ? The thunder and the lightning are the causes of the evil, but not the evil itself.—*French Translator.*

should be weaving in the large laboratory when the early ones have finished their cocoons.

4th. When the silk-worm has cast out the down which precedes the silk, and it has just begun to wind itself in silk, as the air does not then directly strike upon them, the care of the interior circulation need not be so strictly attended to; and the air may be freely admitted now and then, and even when it is agitated.

5th. When the cocoon has acquired a certain consistency, the laboratory may be left quite open, without fearing the variations of the atmosphere. The tissue of the cocoon is so close, that the agitation of the air, far from being detrimental to the silk-worms, agrees with them, even if it should be colder than the temperature fixed for the laboratory.

What I have been stating shews the great advantage of having all the worms equal in the laboratory, and that they rise at one time as nearly as possible. For were there a great disproportion of time among them, the general rearing would not proceed well, and the loss would be very great.

In countries where, by the effects of the climate, the temperature is hotter than that which I have stated for the period of rising, the air is dry without being much agitated as it is in more tem-

perate regions, particularly near mountains; therefore, in those climates it is only necessary to leave
the current of air free on the side where it blows
coolest.

Although it may seem needless, to those who
inhabit warm climates, that I should enter into
such minute details as those I have offered ;—as in
an elementary work, rules should be laid down
applicable to all cases and to all places, in the art
of which it treats ;—I have endeavoured to speak
of every circumstance that might occur, and to
provide for it.   All the care I have hitherto recommended has tended,

1st. To preserve the silk contained in the reservoirs of the silk-worms in a constantly fluid
state.

2d. To keep the skin or surface of the silk-
worm sufficiently dry, and constantly in the
degree of contraction necessary, and without
which the silk-worm would perish.

3d. To prevent the air from ever being corrupt,
and which might make the silk-worm ill, or suffo-
cate it, at those very periods when it most needs
its highest vigour to pour out all the silk it
contains.

If these rules are not observed with exactitude,
there is danger of the accidents occurring which
it may be useful here to state.

1st. Too cold or agitated an air, introduced into

the laboratory, may instantly harden, more or less, the silky substance of those worms on which it may blow. This substance thus not being fit to pass through the silk-spinning tubes, the insect is soon obliged to cease drawing out its cocoon, and suffers. Then will many of those worms that are not sufficiently wrapt in the silk be liable to drop off at any moment, and lessen the abundance of cocoons. To be convinced of this, it is only necessary to make the following experiment:—Cover several small fagots with paper, when they are loaded with cocoons, being careful to place the paper only between the second and third fagot, the fourth and fifth, and the sixth and seventh, and so on: doing this on the side exposed to the blast and agitation of the air, it will be found, that on those fagots that are sheltered from the air by the paper, the cocoons will be fine, full, and numerous; whilst on the exposed fagots there will be few cocoons, the worms having dropped off, gone elsewhere, or formed bad cocoons.

2nd. Too damp an atmosphere, preventing the contraction of the skin of the worm,—which enables it to evacuate the last excrement, and to exude the silk through the silk-drawing tubes,—causes them to suffer, weakens them, slackens their work, and gives them numerous disorders which cannot easily be defined.

3. An atmosphere vitiated by the fermentation of leaves and dirt, or by the later worms that lie on the litter, as well as by the defect of circulation in the interior air, which renders the breathing of these insects difficult, relaxes their organs, and also causes various diseases among them. In such cases, many worms drop off, others form bad cocoons, die within them when they are finished, and are spoiled.

4. A case of very rare occurrence here, but which I shall note, to complete my views on this subject, is, too warm and dry an atmosphere, which dries up the worms, producing too violent a contraction of the skin, not proportioned to the vacuum which increases in the animal by the slow pouring out of the silky substance, and by transpiration; and thus forces them to violent and fatiguing action in the formation of the cocoon. In which case they empty the reservoirs of silk too fast, forcing the silk-drawing tubes, producing coarser silk, which thus never can have that fineness which it possesses when produced in a temperature of 69°. Having tried to expose a number of silk-worms to very dry air, at 100° of temperature, I obtained from the cocoons, by the common method of spinning several thousand feet of the coarse downy floss or *bave,* the weight of this floss being six times

K

greater than the floss obtained from cocoons formed in a temperature of 69°. This observation may explain why the silk produced in very hot climates is stronger and less fine than that produced in temperate regions, where the silk-worms are reared at a lower degree of temperature.

Art teaches us to avoid all the inconveniences of which we have spoken; inconveniences which, every year, destroy an enormous number of silk-worms, and tend to form a large proportion of defective cocoons.

There are general and confused notions upon the diseases of silk-worms; but as the causes of those diseases have not been sufficiently investigated, it often happens that instead of a remedy, a poison is administered.

The hygrometer and thermometer having shewn the causes of disease, also supply the means of curing them, and the means are those which I have recommended, fires in the grates in the angles of the laboratory, light-blazing fires now and then when required; opening the ventilators, fumigating bottles, &c. &c.

The fifth age is accomplished when the silk-worm pours out its silk, and forms the cocoon.

The fifth age is perfected, when on touching the cocoon, it appears to have attained a certain consistency; the silk-worm has then cast its envelope, is changed into the chrysalis, and has entered its sixth age.

7. *Quantity of Vapour, Gas, and Excremental Substance, emitted by the Silk-worm, from the time it reached its highest point of Growth, until it has reached maturity, and until the perfect formation of the Cocoon.*

I here offer the calculation resulting from facts, by which I have been able to ascertain the quantity of substance which issues from the silk-worm towards the close of the fifth age : that this calculation may shew the evils which are constantly likely to attack a laboratory.

It must be well observed, that I only allude to the noxious emanations exhaled by the silk-worms, and not of the leaves, fibrous fragments, and excrements, all which substances deteriorate the air, and are injurious to the silk-worms, if not speedily removed; of these I shall speak separately. (Chap. XIV.)

The result of my experiments proves, that 360 silkworms, which produce about one pound and a half of cocoons, weigh, when at their highest growth and size, three pounds, three ounces and a half.

The silk-worms, after this, are ready to begin their cocoons, in the course of two or three days, and then only weigh about two pounds, seven ounces.

When the silk-worms begin to rise, they void a quantity of nearly pure water, part of which is sometimes discharged through the silk-drawing tubes, and by transpiration. They also evacuate a small quantity of solid substance, and then form the cocoon in three or four days; these cocoons altogether weigh about one pound and a half.

Let us now imagine a laboratory similar to that which I have hitherto described, calculated to contain silk-worms proceeding from five ounces of eggs, and sufficient to produce about six quintals of cocoons; the following will offer the result:—

1st. If 360 silk-worms weigh three pounds, three ounces and a half, when in their utmost growth and perfection, it must clearly appear that the whole of the silk-worms of the laboratory, which produce 600 pounds weight of cocoons, will weigh 1285 pounds, three ounces, when they reach their utmost growth.

And if the 360 silk-worms, previous to beginning their cocoon, only weigh 42 ounces, it must appear equally clear, that the whole of the silk-worms of the laboratory will be about 10 quintals, 50 pounds: and therefore, in three days the silk-worms must have lost 237$\frac{1}{4}$ pounds weight of substance, either solid or liquid, from exhalation, or steam.

2nd. And if after two or three days, the silk-worms, that are reduced to 10 quintals and 50 pounds weight, are changed into 600 pounds weight of cocoons, it is evident, that in three or four days they must have lost 450 pounds weight of substance, either in liquid, or in vapour and gas.

3rd. In the space of six or seven days, therefore, the bodies of the insects requisite to produce only 600 pounds of cocoons, must have lost 700 pounds weight of vapour or gas, solid and liquid excremental substance ; this astonishing quantity of substance, excreted from the bodies of the silk-worms in so short a time, is of greater weight than the total weight of the cocoons and aurelias, which only weigh 600 pounds. It is scarcely credible that the bodies of the silk-worms should yield so much noxious matter in a few days, were it not demonstrated by positive facts. (Chap. XIV.)

It is needless to remark how much this large body of exhalation, were it stagnant in the laboratory, might, in the latter days, generate disorders and disease most quickly, and cause great mortality at the very moment when the abundant crop of the cocoons was most confidently expected. We must, therefore, feel the deep necessity of attentively following the prescribed directions for avoiding this evil.

Chapter IX.

OF THE SIXTH AGE OF THE SILK-WORMS, OR OF
THE CHRYSALIS, GATHERING, PRESERVATION,
AND DIMINUTION IN THE WEIGHT OF THE CO-
COON.

WE have seen in the preceding chapters, that the
fifth age of the silk-worms commences after the
fourth casting of the skin, and ends when the insect has commenced its cocoon, and is transformed
into the chrysalis, leaving their former envelope in
the cocoon.

The sixth age begins in the chrysalis state, and
ends when they appear as moths, having left their
shell in the cocoon that covered them.

This age requires less care and attention than
those preceding it, particularly if all the previous
directions have been minutely adhered to.

Still are the various operations required during
this age of some importance, nor are they quite
uninteresting. The following are the necessary
things that remain to be done.

1st. To gather the cocoons.

2nd. To choose the cocoons which are to be
preserved for the eggs or seed.

3rd. Preservation of the cocoons until the appearance of the moth.

4th. Daily loss of weight which the cocoons

suffer from the time they are finished, until the appearance of the moth.

1. *Gathering of the Cocoons.*

According to my experiments, strong, healthy, and well-managed· silk-worms will complete the cocoon in three days and a half at farthest, reckoning, from the moment when they first begin casting the floss.

This period will be shorter if the silk-worms spin the silk in a higher temperature than that which I had fixed, and in very dry air.

It is also more or less prolonged if the silk-worms are not well and healthy, or if they are exposed to a colder temperature than has been fixed. If they are exposed to transitions of heat and cold, to damp and vitiated air, or to draughts of wind before the cocoon is sufficiently advanced to shelter them entirely; and in short, if a great number of the silk-worms rise long after the first have risen, which is always the consequence of bad management and ill-directed care.

I allow it may appear difficult to old practical cultivators immediately to change their old habits to adopt new customs, however easy they may prove.

To avoid the losses which any slight inattention may have occasioned, it will be better they should not take off the cocoons before the eighth or ninth day, reckoning from the time when the silk-worms first rose. I take them off the seventh, and even the sixth day, because my laboratories are conducted with such regularity that I know with certainty when I may do so.

We shall see besides, that this delay only occasions a trifling diminution in the weight of the cocoon, rather improves the quality of them, and particularly the cocoons of which the worms having taken a longer time to purge themselves, consequently are slower in completing the cocoon.

When the seven or eight days are elapsed, the cocoon should be gathered.

This operation should begin on the lower tier of hurdles, removing all the cocoons regularly from the hedges and hurdles, and where there are no branches.

The fagots should not be flung down, but gently taken off the hurdles, and given to those who are to gather the cocoons.

By flinging down the fagots as some do, there is a chance of crushing some of the cocoons, and staining them by killing the worm, before the work is well finished.

No such accidents can occur in a well-managed laboratory.

The persons directed to gather the cocoons should be seated in rows, ready to strip the loaded fagots, which should be brought to them, and laid near them gently on the floor.

There should be a basket placed between two of the gatherers to receive the cocoons, and another person should be employed in removing the stripped fagots, which fagots, if they are composed of heath or broom, may be laid by for another year; but if they are made of straw, turnip haulm, or any such light material, are not worth saving, and may be burnt*.

* Heath may last for several years, and is better for the purpose the second than the first year.

Straw, when used to form the hedges, and indeed heath, the first year, is apt to injure the silk-worms, when the ends and extremities are too slender and thin.

For when the silk-worm rises on the small boughs, it always climbs up to the very ends, and if these ends are too slight, they bend and part from the stronger twigs, not having sufficient support; the silk-worm drops on the hurdle, and on the ground, if the branches have not been arranged so as to avoid this. These falls are very injurious, and sometimes kill the silk-worm; therefore the heath should be carefully placed to prevent these falls.

When the heath has been used one year, I have the bundle or fagot just passed through a light blazing fire, which singes off any down or floss that may remain on the boughs and the ends of the leaves, and renders the heath perfect for the use of the next year. When singed, they should be well beaten out against a wall, to shake out the burnt particles, and aired, that they may retain no smell.

They may then be stowed away, till wanted, when they should again be well aired. Instead of old dry heath, it would be an advantage to use the branches of fresh plants, such as turnip plants, but they should be stripped of all the leaves, and of the small weak top twigs.

K 5

The persons who gather the cocoons should have a sheet of paper before them, to avoid dirting themselves with any worms that may be spoilt.

All the cocoons that want a certain consistency, and feel soft to the touch, should be laid aside; this must be rigorously enforced, that purchasers may not have any opportunity of lowering the price of the cocoons they want to buy, under the pretext that they are of a mixed and inferior quality. There is another advantage which may be derived from strictness on this point. I, every year, made very near as much by the silk I had spun from these refuse cocoons, as by the sale of the finest cocoons; the silk may also be used for domestic purposes.

But to return to my subject; the cocoons should be emptied out of the baskets upon hurdles or trays, placed in rows, and raised a certain distance from the ground, that they may be conveniently sorted and examined.

The cocoons should be spread out about four fingers deep, or nearly up to the top of the edge of the wicker tray.

The baskets, the floor, and all things used for gathering or holding the cocoons, should be most carefully cleaned.

The labour of those employed within the laboratory should be so directed as to keep time with the work of those without, employed in picking the cocoons from the bushes

When the cocoons are detached from the bushes, the down, or floss, in which the silk-worms have formed the cocoon, should be lightly and quickly taken off.

Beginning at sun-rise, and working till four o'clock P.M., twelve persons will be sufficient to gather 600 pounds weight of cocoons from the bushes, clean them, and spread them on the wicker hurdles.

As soon as this is done, the cocoons may be put into baskets; if they are for sale, they must be weighed and sent off to the purchaser. Before they are removed the sheets of paper, walls, wood-work, and all holes and corners, should be well searched, in which the silk-worms can have made their cocoons ; if any are found, they should be taken off, and well cleaned, before they are mixed with the rest, that they may all be equally fine.

It will always appear that the quantity or weight of the cocoons is in exact proportion to the space of the wicker hurdle which the silk-worms have occupied.

Following the method I have recommended, from 183 feet 4 inches square of wicker, upon which have been reared the silk-worms proceeding from one ounce of eggs, these will produce about 112 to 127 pounds of cocoons of the very best quality.

Whether the laboratory be large or small, the produce will always be exactly in the above-stated proportion, and will not diminish, let the season be ever so unfavourable, if the rules I have prescribed are strictly adhered to.

2. *Choosing the Cocoons for the Production of the Eggs.*

In the present imperfect state of the art of rearing silk-worms, it requires at least, to save the six-tieth portion of the cocoons that are gathered.

This calculation is founded on a series of experiments which tend to demonstrate it.

1st. About two ounces of eggs may be saved out of one pound and a half of male and female cocoons *.

2d. Generally in different parts of Italy, where silk-worms are reared, they only obtain 45 pounds of cocoons from an ounce of eggs ; and as it is a fact, that in the whole country that used to form the kingdom of Italy, the value of export of silk, and the whole produce of the cocoons, amount to above 80 millions ; it must appear evi-

---

* The author of the article on silk-worms, in the Cours d'Agriculture of L'Abbé Rozier, says, that the common average is one ounce of eggs from one pound of cocoons. The superior difference in the quantity of seed obtained by Count Dandolo, is, no doubt, owing to the improved management of his silk-worms.—*(Translator.)*

dent that the value of the cocoons used for the production of the eggs, amount to near a million and a half, which is thus taken from our foreign trade.

If then, by perfecting the art of cultivating silkworms, which is my present object, the production of cocoons is increased from 45 pounds to 90 pounds weight from one ounce of eggs; it is clearly shewn that we shall add to our exports a quantity of silk equal to half the number of cocoons employed for the production of the eggs, which produces a considerable sum.

This advantage is one of the least which the improvement of the care of silk-worms may offer, as I shall shortly prove. (Chap. XV.)

To return to the cocoons destined for seed, it may be said with security, that if they are taken from a well-managed laboratory, it is needless to give oneself much trouble in choosing and sorting them. Several experiments have proved this, and various persons who took some of my cocoons without particular choice, always obtained good eggs from them.

However, in the present state of the management of silk-worms, it would be striking too suddenly at the prejudices of the cultivators, to desire them to suppress this picking and choosing of the cocoon, although it be a mere waste of time; particularly as every accident that occurred in the course of the rearing of the silkworm would be attributed to the omission of this custom of choosing. Time and experience will convince those who know their own interest, of the inutility of choosing the cocoon.

If this choice is to be made, the straw-coloured cocoon should be preferred, the hardest, particularly when the two extremities are hard, and the web fine ; those that are a little depressed in the middle, as if tightened by a ring, or circle, and not the largest.

The small cocoons, peculiarly hard at the ends, and a little depressed in the middle, shew that the worm had much strength, since it fastened the floss firmly, and rounded it well, working the extremities thoroughly, which a weak worm could not have done.

Hitherto I have not been able to discover by any experiment that the strength shewn by the silk-worm in the formation of the cocoon had any influence upon the fecundity of the male, or upon the quality of the eggs in the female. Cocoons, of various tenuity and different shapes, have equally afforded me large quantities of well-impregnated eggs. Healthy silk-worms, perfectly mature, of equal weight, have given cocoons that varied in weight.

It is a fact, that the greater quantity of silky substance drawn from a healthy silk-worm, than that afforded by a worm equally healthy, only demonstrates, that one had accumulated a greater abundance of silk in its reservoirs than the other, without inducing us to think that they differ in fecundating strength.

The perfect health of the silk-worm depends not on the greater or less quantity of silky substance which it can produce. A silk-worm may be strong and healthy, and yet contain less silk than a worm of a weaker appearance.

I have found some sick silk-worms, that would not even have had strength to form a cocoon at all, if I had not assisted them, and were contracted and swelled, and are termed *harpions*, (or *riccioni*) *. I have found in some of these more silky substance, than in many perfectly healthy worms that I opened for examination. (Chap. XII).

The result of the experiments I made on some of my own cocoons, and on those of my tenants, that from a cocoon formed by a healthy silk-worm, there will always issue a very good moth, either male or female, and I admit no exceptions to this rule.

When choosing cocoons for seed, some persons will shake them one after another, to find out whether the chrysalis rattles in the cocoon, which they consider a sign of a healthy chrysalis; this

* Of all the diseases of silk-worms described by Count Dandolo, the *Riccione* is the only one which is to be found described in the Cours d'Agriculture of L'Abbé Rozier, under the name of *Harpion,* or *Passis.*

is tedious and unnecessary *. The chrysalis always exists in the cocoon, and is sure to be healthy, if the laboratories have been well managed.

It will happen sometimes that the silk-worms in forming the cocoon, when drawing the last portion of the silk, cannot carry the floss quite perfectly from one extremity to the other, or do not fasten it well to the inside; notwithstanding this, which prevents the extremities being well covered with silk, and causes the threads to cross the cocoon irregularly, the chrysalis will be found to be perfect and healthy.

In such instances, on shaking the cocoon, it may very probably not rattle, being fixed by the threads, although it be perfectly sound. However, such as are inclined to make these trials may do so; I have stated my opinion of their inutility.

There are no certain signs to distinguish the

* The following observation is from L'Abbé Rozier's work: " When choosing the cocoons for the production of the moths by shaking the cocoon close to the ear, may be ascertained whether the chrysalis be alive. If it is dead, and loosened from the cocoon, it yields a sharp sound. The *muscardin*, or *cocon dragée* [a], produces the same sound. But when the chrysalis is alive, it yields a dumb muffled sound, and is more confined in the cocoon.—(*French Translator.*)

[a] There are no corresponding terms in English for the various diseases of Silk-worms, and it has been thought best to retain the original Italian or French term.—(*English Translator.*)

cocoons which are to produce the male moth from those that contain the female; but the least erroneous, and best known, are the following :—

The cocoon is smaller, sharper at one or both ends, and depressed in the middle, which generally produces the male; the round full cocoon, without ring or depression in the middle, usually contains the female.

In the following Chapter, we shall observe that the chrysalis and female cocoons are nearly twice as heavy as the male; which must naturally presuppose that the female cocoon, all circumstances equal, must be larger than the male.

Having composed tables of cocoons which I believed to be all male, and others of female cocoons, I found that in both cases the greatest portion verified the signs I have described as likely to distinguish them.

A silk-worm, although female, will frequently form a small cocoon, sharp at the ends ; because, having had great vigour, it has moved, and turned, and composed it with great ease and freedom ; whilst, on the contrary, when the insect is weak and languid, its motions are less powerful, which will often cause a male worm to form a large thick cocoon without any sharpness at the extremities.

We must, therefore, conclude that the cocoons proceeding from well-managed laboratories, which are solid and fine-grained, are all likely to produce good eggs ; in a hundred, there will scarcely be one that will fail in producing a strong sound moth.

We must also add, that as to the means of distinguishing the sexes of the moths, though there may be signs that give general indications of the difference, yet they are not so infallible as not sometimes to mislead us.

3. *Preservation of the Cocoon intended for Seed.*

The preservation of the cocoons destined to produce eggs, is an operation of some importance.

It will require a very dry room, exposed to a temperature of between 66° and 73°.

Experience shews us, that were the temperature above 73°, the transition of the chrysalis to the moth state would be too rapid, and the coupling will not be so productive. If the temperature is below 66°, the development of the moth is tardy, which is also injurious, as we shall shew hereafter. If the apartment is not dry, the damp, which was hurtful to the worms, will be equally so to the chrysalis, and change it into a weak and sickly moth. The apartment should be kept in an even temperature, between 66° and 73°, and the rooms on the first floor should be preferred to a ground-floor apartment.

As soon as the cocoons are collected which are intended to produce eggs, and are spread either on a dry floor or on tables ; an active hand should strip them of any down or floss that may still hang about them.

This floss does not form a part of the cocoon, and must be removed, because the cocoon is cleaner, and not so apt to get soiled without it; and also that the moth may not get its feet entangled in the floss when it first appears. I have known them sometimes obliged to have assistance to get rid of the floss. This is a very tedious part of the work; however, a good hand at it will clean thirty pounds a day, without much exertion. While cleaning the cocoons, all those that appear to have any flaw or defect should be laid aside; this is also the time to separate the male and female cocoons, as far as we can distinguish them.

When this is done, the sorted cocoons must be put on the tables, in layers of about two inches, allowing the air to pass freely through them, that it may not be necessary to stir them too frequently.

When they are too much heaped up, they must be constantly turned and stirred; and as the chrysalis produces a constant evaporation, as will be seen hereafter, it follows that when the undermost cocoons are not stirred, they may become moist, and injure the chrysalis.

If the heat of the apartment disposed for this purpose is above 73°, and that the cocoons cannot be placed elsewhere, every method of diminishing the heat should be tried; such as keeping all apertures to the sunny side carefully closed, to establish thorough draughts of air, to dry the humidity which exhales from the chrysalides.

It is also beneficial to stir the cocoons round once a day, even when they are thinly spread out, if the atmosphere continues loaded with moisture; but should the temperature rise to 78° or 82°, the cocoons must, without delay, be put into a cooler place. Moderate temperatures are, without exception, best adapted to the silk-worm, the chrysalis, and the moth.

4. *Daily loss of weight suffered by the Cocoon from the time of its Formation, till the Moth escapes from it.*

No branch of information, however minute, can be useless when it can in any degree contribute to improvement, when it diminishes losses, and when it increases the profits of any art whatever; and as my object is to enable any body to rear silk-worms, and to draw from them every possible advantage they can offer, I have even tried to ascertain the exact loss of weight of the cocoon each day.

It is a common opinion, that for a certain period the cocoon diminishes in weight, and then increases. This old error induces several persons to give the cocoons too soon to the spinner, before they lose weight; or too late, when they keep them back, in hope they will soon recover weight.

I cannot trace the origin of this error. Can it have been suggested by the silk-spinners for their own interest?

To enable me exactly to know and calculate the diminution of the weight in the cocoon, I carefully weighed, every day, 1000 ounces of cocoons, reckoning from the moment they completed their formation, until I perceived that some moths, by wetting the cocoon a little, indicated that they had pierced the envelope which covered the chrysalis, and were preparing to rend the cocoon.

The following is the result of the daily decrease of the 1000 cocoons, in a temperature between 71° and 73°:—

	Ounces.
Gathered from the fagots and cleaned, the cocoons weighed	1000
First day following, the cocoons weighed . .	991
Second day	982
Third day	975
Fourth day	970
Fifth day	966
Sixth day	960
Seventh day	952
Eighth day	943
Ninth day	934
Tenth day	925

We find by this that the cocoons lose, in ten days, seven and a half per cent., by the desiccation of the chrysalis alone. The first four days they lose three per cent., or three-quarters per cent. a day; in the last days they lose rather more, because as the formation of the moth approaches, a greater quantity of humidity evaporates.

The drought or dampness of the atmosphere may increase or diminish the loss by some ounces.

It is therefore evident that those who, to please the silk-spinners, leave the cocoon on the fagots, lose a certain sum on every pound of cocoons which they sell.

The proprietors whose silk-worms have risen at various times some five or six days later than others, and who have not gathered the cocoons until the twelfth or thirteenth day after the earliest rose upon the hedges, are likely to suffer losses of three or four per cent., without any chance of reaping any possible advantage from the sacrifice.

In most cases, it is a loss for the purchaser of the cocoon, who has a view to spinning the silk, to receive those that are of different ages; because when in some cocoons the moth is preparing to come forth, and other cocoons are not so forward, the spinners are at a loss whether to let it come out directly, or to kill the chrysalis to preserve the cocoon.

If the rules I have recommended in the preceding Chapter are exactly followed, this loss will be avoided; and the cocoons will be perfectly formed, and ready to be worked off, at the end of seven days, reckoning from the day they first rose upon the hedges.

CHAPTER X.

OF THE SEVENTH AGE OF THE SILK-WORM; OF
THE BIRTH AND COUPLING OF THE MOTH;
OF LAYING THE EGGS, AND THE PRESER-
VATION OF THE EGGS.

THE seventh and last age of the silk-worm comprises the entire life of the moth.

It is not in a work of this kind that I need demonstrate how, within the envelope which covered
the chrysalis, the moth is formed by its powerful
vitality and chemical affinities; and also how is
formed the fecundating substance, and a particular kind of fluid matter which accumulates in its
various reservoirs, and, in short, every part of its
being.

I will only say, that when the moth is formed it
pours from its mouth a tasteless liquid, nearly
like that which moistens and softens the envelope
which wraps it, and also the strong web of the
cocoon in which it is inclosed.

The formation of the moth, and its disposition
to issue from the cocoon, may be ascertained when
one of the extremities of the cocoon is perceived
to be wet, which is the part occupied by the head
of the moth. Some hours after these signs appear,
and sometimes even in one hour after, the moth
will pierce through the cocoon, and come out.

It sometimes will occur that the cocoon is so hard, and so wound in silk, that the moth in vain strives to come forth, and dies in the cocoon.

Sometimes the female deposits some eggs in the cocoon before she can get out, and often perishes in it.

May not this observation shew us the use of extracting the chrysalis from the cocoon, by cutting it, that the moth may only have to pierce its envelope? I myself have practised it successfully; but I found the operation so tedious, that when to this is added the disadvantage which it is to the moth not to find its cocoon to stretch itself upon where it first comes forth, I do not advise the opening of the cocoon by artificial means *.

* It is very favourable to the moths, when they put forth their head and first legs, to find some substance to which they may fasten, and thus facilitate clearing out of the cocoon by the support; for this they should be put in layers two inches deep.

If the chrysalis is drawn from the cocoon to facilitate the moth's issue from its envelope, it will occur that should the moths be put on a smooth surface, five in a hundred will not be able to get out, but drag the envelope along, and at last die, not being able to disencumber themselves.

If the surface on which the chrysalis is placed is not planed, the moths will issue with greater ease, as the inequalities of the surface assist them by supporting them.

I therefore imagine the method I first described is altogether the most practical, being careful to put the cocoons at equal distances as fast as the moths appear, and to remove the cocoons that are pierced.

The life of the moth lasts ten, eleven, or twelve days, according to the strength of its constitution and the mildness of the atmosphere. A hot temperature tends to accelerate all the operations which nature has destined this insect to execute, and also accelerates the desiccation which precedes its death.

This last age requires attentive care. Although the moth of the silk-worm has wings similar to those of the common race of butterflies, they possess not strength to fly and seek a shelter for the eggs they wish to deposit, as the flies of all other caterpillars do. (Chap. I.)

It therefore depends on the industry of man to collect and preserve the eggs of the silk-worms, and to dispose of them in the most advantageous manner for the following year.

It appears to me to be for the common interest of those who rear silk-worms to obtain from their own cocoons good eggs, rather than to buy them, that they may feel sure of the excellence of the eggs. However, few people do this. I will therefore shew the ease with which this may be effected, and the simple and sure means of obtaining, for a small sum, a large quantity of the best eggs.

I should imagine there can be but three motives which prevent cultivators from preserving the

L

eggs of their own gathering, and induce them to prefer buying them from others.

The first may be, that the broods have failed, and have produced bad cocoons : this motive never could exist, were the silk-worms well managed.

As to the second motive, experience constantly demonstrates, that eggs that are gathered and appear to proceed from good cocoons fail, and are not so successful as the produce of eggs that are bought, it will only clearly prove that the silk-worms of the buyer are worse managed than those of him who sells.

The third motive is, that, to save themselves trouble, cultivators will rather buy eggs than rear them, so as they can believe them to be of a tolerably good quality, and produced from well-managed silk-worms; which shews that laziness is the great inducement to buy eggs, instead of rearing them.

The cases that can make it necessary to purchase eggs, sooner than to raise them, are therefore of very rare occurrence.

I should also add, in this Chapter, that there exists a notion that every two, or three, or four years, the eggs proceeding from a laboratory should be changed; it requires but little to be said on these egregious popular errors.

If for a thousand years a laboratory produced good cocoons, the eggs of those cocoons, being carefully preserved, will for a thousand years continue good and undegenerate, like the eggs of any other oviparous domestic animal that we are acquainted with.

To suppose that the good cocoons of a cultivator, after a few years, are no longer fit to produce good seed, and yet that those cocoons can give good seed for the use of any other laboratory, would be to admit a superstitious contradiction which reason, practice, and science alike condemn.

We shall comprise in three paragraphs all that treats on the production and preservation of the eggs :—

1st. Birth and coupling of the moth.

2nd. Separation of the moth, and deposition of the impregnated eggs.

3rd. Preservation of the eggs.

1. *Hatching of the Moths, and their Propagation.*

If the cocoons that have been selected to produce eggs are kept in a temperature of 66°, the moths begin to be hatched after fifteen days ; if the cocoons are kept in a heat between 71° and 73°, they begin to come forth after eleven or twelve days.

In the first case, all the moths occupy about

fourteen or fifteen days in hatching ; in the latter, they employ only about eleven or twelve.

This law is for the most part general, though there are some exceptions. As I have said before, a sign that the moths are about to come forth is, when the cocoons are humid or wet at the end where the head of the moth is situated.

The room in which the moths are produced should be dark, or at least there ought to be only sufficient light to enable one to distinguish objects.

The moths do not come forth in a great number the first nor the second day; they are hatched chiefly on the fourth, fifth, sixth, and seventh days, according to the degree of heat of the place in which the cocoons are kept.

The hours when the moths burst the cocoon in greatest number are the three or four hours after sunrise; very few are produced during the other hours of the day, if the temperature be from 64° to 66°; though, if it be kept at 73°, more are hatched. During the days when the most are hatched, the surface of the cocoons is seen nearly covered with them. Some persons think that the first which come forth are males ; for my own part, I have observed both males and females, nor do I think there is any thing fixed on that point.

The male moths, the very moment they burst the cocoon, go eagerly in quest of the female.

I have observed, in another place, that the male cocoons are very difficultly to be distinguished from the female; though there are, however, some marks which enable one to discriminate a good number. (Chap. IX. § 2.)

Nevertheless, it is always very advantageous to separate those cocoons which appear to be male from the others. By this means fewer unite on the tables, and consequently,—

1st. They are observed in succession, and those which are united can be removed.

2nd. Those not united can be allowed to remain longer on the table, which is productive of advantage, as we shall see afterwards.

3rd. It is more easy afterwards to bring them together, as it is more easy to take up those which are single than those which are united.

The following is the best method to facilitate the hatching and subsequent union of the moths:—

As I have said before, the moths begin to burst the cocoon as soon as it is daylight; but they do not come forth in such great numbers during the first or second hour, as during the third and fourth.

When the moths are seen united, they must be placed on a kind of frames covered with linen (Fig. 26.), expressly made in such a manner as to allow the linen to be changed when it is dirty.

The perfect union of the sexes is indicated by the fluttering of the male.

Much care must be taken in raising the united moths. They must be held by the wings, in order not to separate them; and if this happens, they must be replaced on the tables of the moths of their own sex.

When one small table is filled with moths in a state of union, they are to be carried into a moderately-sized room, sufficiently airy and fresh, and which can be made very dark. These little tables are to be placed on the ground, or elsewhere.

Having employed the first hours of the day in selecting and carrying the united moths, the males and females which are found separate on the tables are to be brought into contact.

This operation, though tedious, is not difficult; the males and females are alternately raised, and put together on other frames, which are carried into the dark room.

At the end of a certain time, it is easy to be ascertained if there are more females than males. The female is easily distinguished by the greater size of its stomach, which is almost as big again as that of the male.

I have also tried this by their weight, 100 males weighing 1700 grains, while 100 females weighed

3000 grains. It is therefore unnecessary to point out other characters to distinguish the males from the females : besides, the male which is single beats about its wings at the approach of the least light.

For reasons which I shall afterwards point out, the hour must be noted at which the tables containing the united moths were placed in the dark chamber. The same ought to be done with respect to the other little tables with the moths that afterwards unite.

If, after this operation is over, there still remain some moths of each sex, they are to be placed in the small perforated box, (Fig. 27.) until the moment favourable for their union comes.

From time to time, they must be looked at, to see if they separate, in order that they may be brought anew into contact. When any thing is to be done in the dark chamber, as little light as possible must be admitted, only sufficient to distinguish objects. The more light there is, the more are the moths disturbed and troubled in their operations, as light is too stimulating for them.

The moth of the silk-worm belongs to that kind that fly by night, and which we often see flitting round lighted candles ; where they are called *Phalènes*, or night-butterflies, to distinguish them from those which fly by day, called in consequence day-butterflies. The boxes (Fig. 27.)

are very convenient, particularly to keep quiet
the males which remain.   It is difficult, however,
to prevent the male moths from striking about
with their wings.   When they make this motion,
there is detached from their wings a sort of down,
which makes much dust, which sticks to every
place, and incommodes breathing.   If care be
not taken to quiet this motion by darkness, there
will ensue an almost entire destruction of their
wings, and consequently a great loss of their vital
powers.

While employed in carrying off the united
moths, and while others are coming forth, care
must be taken to remove the cocoons that are
burst ; as these cocoons are moist, they communicate their humidity to those which are still
entire.

The paper, also, that is on the wicker trays, is
easily soiled ; these dirtied portions must be
changed, in order that the trays and cocoons
should be kept as clean as possible, to prevent
the air of the room from being corrupted.   During
hot weather, constant attention is required during
the whole of the day, as there is a constant succession in the processes of hatching and union
of the moths, which occasionally vary in relative
proportion to one another.

Amongst all the methods which are adopted in
these operations, I recommend that which I have

just explained, as being the most simple, the most easily executed, and as offering the following practical advantages.

1st. The moths being hatched, and remaining almost all of them separate for a short time before they unite, have leisure to evaporate a portion of the humid and earthy matters that load them.

2nd. All those which unite of themselves on the table, are handled only once, when they are raised up ; afterwards they remain quiet, the whole of the time during which they ought to be united.

3d. The moths that are not united, are, by this method, only handled once.

4th. The females and males which remain separate on the tables when the junctions take place, and which have been put into the box, (Fig. 27.) are handled again only when the moths of the required sex are found.

It might seem, by this method, that the cocoons would be much soiled on the wicker trays, but this is not the case. If care be taken frequently to remove the cocoons that are burst, and to stir those which are not yet open, the paper which covers the trays soaks up almost all the moisture of the cocoons that lie upon it ; so that attention being had to change this paper when it is very wet, the cocoons become very little dirty.

You may employ, instead of a frame, paper, pasteboard, or any thing else, for the purpose of receiving the deposited eggs. I speak of **frames**, because they make a part of the description of the utensils necessary in the art of raising silk-worms.

There are very few good cocoons that do not produce a moth, and of these few the greater number consists of those whose hardness and smallness prevent the moth making a hole, by which to come forth. The proportion of weight between the cocoon that still contains the moth, and the empty one, but which is not yet perfectly clean, is as six to one, that is to say, from 28 ounces of full cocoons 4 ounces can be obtained; but from those which are burst, only three-quarters of an ounce. (Chap. XIV.) The relative weight of the two envelopes found in the open cocoon, and the open cocoon itself that has been well cleaned, is about as one to thirteen, or the two envelopes weigh in general half a grain, and the empty cocoon nearly six grains and a half.

2. *Separation of the Moths, and Laying of the Eggs.*

In the preceding paragraph I have supposed, when speaking of the union of the moths, that the number of males was equal to that of females, and that consequently, at the time of their separation, it would only be necessary to keep the females, and throw away the males.

However, that is never the case, as there is always an excess either of males or females.

If there are more males they must be thrown away; but if there be an excess of females, males must be allotted them, that have already been in a state of union. Great care must be taken, when you separate the couples, not to injure the males.

I have observed above, that it is useful to mark the hour when the couplings take place, because the male ought not to remain united more than six hours. After the lapse of that time, you take the two moths by the wings and the body, and separate them gently, which is easily done. All the males which are no longer in union must be placed upon the frames; the most vigorous are afterwards selected, and united with those females which have not yet found a mate. If at the time, more males in a vigorous state are found, than are required, and you foresee that they may be useful afterwards, they must be preserved in a spare box, and kept in darkness. When I am aware that I shall be in want of males, I allow them to remain united with the female, the first time, five hours only instead of six.

It seems that the females are not injured by

waiting for the male, even many hours; the only
mischief that occurs being the loss of some eggs
which are not impregnated.

To preserve the males in a state of vigour till
the moment of coupling comes, they must not be
allowed to beat about their wings. Before separating the two sexes you must prepare, in a cool,
dry, and airy chamber, the linen on which the
moth is to deposit its eggs.

Twenty-two square inches of cloth are sufficient
to contain on its surface six or seven ounces of
eggs.

The following is the manner in which matters
should be arranged: at the bottom of a *tressel* of
light wood, about four feet seven inches high,
and three feet eight inches long, (Fig. 28.) you
place horizontally, on each side of the length, two
little tables, or boards, arranged in such a way,
that one of their sides shall be nailed to the legs
of the tressel, about five inches and a half high
above the ground, and that the other side of the
board shall be a little higher, and project out-
wards. Upon the tressel a piece of cloth must
be placed, about nine feet two inches long, and
which hangs equally on each side of the tressel.
The two ends of the cloth are intended to cover
the boards below. If the tressel is rather more
than three feet eight inches long, you may place
upon it two cloths, which will offer a surface of

from 18 to 20 square feet; and if they are 22 or 23 inches broad, this superficies can contain more than 60 ounces of eggs. The more perpendicular the lateral parts of the tressel are, the less soiled will be the cloth by the evacuation of the liquid matters that come from the moths.

As many tressels must be arranged as will be required for the quantity of eggs to be collected. And I repeat here, that 28 ounces of cocoons give two ounces more eggs, when the moths which come from them are well chosen. (Chap. XIV.

In thus disposing the moths, they have access to the air on all sides, and can be easily handled, that is placed, and replaced, as occasion may require, on all the points of the cloth.

When every thing has been thus arranged, recollecting that the room should be dry, from which all light should be excluded, except what is necessary to see what you are about; the moths that have been united six hours are to be gently separated, the females to be placed on the frame, and carried on linen to the room where the tressels are, and placed there one after the other, beginning at the top of the tressel, and going downwards. This operation must be continued without intermission, as long as you find females that have been united during the necessary time.

The time must be noted at which the moths

are placed on the cloth, taking care to keep those which are placed afterwards separate, to avoid confusion.

As I have said before, the time when the greatest number of moths are hatched begins about six or seven o'clock in the morning. Consequently the coupling takes place about eight o'clock, and at two o'clock in the afternoon the males ought to be detached, and the females deposited in the place above described. The females that have had a virgin mate must be treated in the same manner as those which have been united with one that had been coupled previously five hours. The females should be left on the cloth 36 or 40 hours, without being touched.

I must here observe, on this subject, that the three following qualities of eggs may be obtained on separate pieces of linen.

1st. The eggs of the females that have been united with virgin mates.

2nd. The eggs of females that have been coupled with males not in the former state.

3rd. Those of females, that in the two cases above mentioned, having already laid during the 36 or 40 hours, still are about to lay again.

As the common opinion is that three different qualities of eggs can thus be obtained, those who think so, ought to place them on separate pieces of linen. I ought, however, to observe, that I admit of no difference among these qualities, and that I firmly believe that all the impregnated eggs obtained by the methods previously described, are always fit to produce good silk-worms, provided they have been well preserved.

The true difference in the qualities consists in the great number of eggs not impregnated, that are found among the sorts that are called inferior *.

* When the third laying takes place, and it is found to contain many yellow unimpregnated eggs, or reddish ones imperfectly impregnated, if it is desirable to know precisely the quantity of impregnated eggs from which worms are to be obtained, that must be done which is pointed out in the note of page 67.

The whole of the eggs placed in the stove-room are weighed; the few silk-worms which are hatched the first, and afterwards the third day, are thrown away; then the eggs remaining are weighed, and adding to this weight the twelfth part, to allow for the evaporation which they have undergone, you will have the weight of those which shall have produced the worms.

If the eggs were in great quantity, and you can take away separately the worms that they may have produced, you may keep also those born on the fourth day, if they were in sufficient quantity. When the weight of the eggs put in the stove-room is exactly known, when you can easily separate the empty shells of those which have produced worms, to weigh those which remain, and when one knows what is to be added to the weight of the eggs that are not yet burst, and how the weight of the worms hatched the first day, and rejected, is to be calculated; it seems to me that nothing more is required to enable us to proceed with the greatest precision.

The moth produces, in the first 36 or 40 hours, the greatest part of the eggs it contains ; those which are afterwards laid, amount only to about the sixth part of those already deposited.   There are, however, some moths which produce more than the sixth part after the first 36 or 40 hours.

Particular differences among the females occasion a great difference in the time employed in laying their eggs.

Of all the different methods employed to obtain eggs, that which I have explained procures the greatest quantity of them.

When, after 36 or 40 hours, you have removed the moths from a part of the linen, it is observed not to be well stocked with eggs, other females must be placed there, in order that the eggs may be equally distributed over the whole of the linen.   Some moths crawl about on the linen, and sometimes stray from it ; however, in general, they remain fixed on the spot where they were placed, or remove very little from it.

When the season, or the temperature of the room, is too hot, that is to say, when it mounts up to 78° or 80°, or when it is too cold, for instance, 64° or 66°, you will find more or less of yellow unimpregnated eggs, or of a reddish colour imperfectly impregnated, which do not produce worms.

Having separated, with care, these eggs from the

impregnated ones, I have found that they have
formed the seventh or eighth part of them.
That was particularly the case in 1813 ; the
temperature was 62°, 64°, 66°, during almost the
whole time of the collection of the cocoons, and
even after that the eggs burst. In a similar case,
the means which I have described above ought to
be adopted, in order to obtain a more suitable temperature. Sometimes it happens also that some
female moth escapes from its mate before impregnation, which produces many unimpregnated
eggs.

Eight or ten days after the deposition of the
eggs, the jonquil-colour, which is peculiar to
them, becomes deeper, then changes into a reddish gray, and afterwards into a pale clay-colour.
All these changes of colour come from the fluid
of the eggs, and not from the shell, which is
almost transparent. (Chap. V.)

Whether the eggs be not at all impregnated,
or only imperfectly so, they are always of a
lenticular form. A little time after their production, there takes place, in the centre of both
surfaces, a depression, which proves that a portion
of the aqueous part of the egg has been disengaged,
and that a kind of drying process has been
effected. There is scarcely any difference of
weight between impregnated eggs. (Chap. V.
§ 3.)

In 15 or 20 days, according to the different degrees of temperature of the rooms, the eggs undergo almost all the gradations of colour mentioned above, and possess then the characters of impregnated eggs.

When all the operations of the 7th age are finished, there is nothing else to attend to, except the preservation of the eggs.

I conclude this paragraph by observing, that in the 7th age, the impregnated female, which weighed then about 30 grains, in three or four days after having deposited its eggs, weighs only about 12 grains. When it is dead and dried up, its weight is only three grains and a half.

3. *Preservation of the Eggs.*

When the eggs have acquired the grayish colour peculiar to them in the impregnated state, and the linen cloths are quite dry, it is time to think of the means of preserving them.

The linen cloths upon which the eggs are deposited may be left there in the same place, provided the heat of the room does not exceed 66° or 68°.

If the temperature of the room is higher, the linen cloths must be removed to a cooler place.

At the extremity of the cloths which cover the shelves of the tressels, eggs are found detached

and fallen down in moving the cloths. They
must be collected into a small pasteboard box;
the layer should not be more than half the
breadth of your finger thick. The same must be
done with all the eggs found attached elsewhere
but on the cloths.

It is of little consequence that all the eggs
should be good. When you wish to hatch them,
they must be weighed as soon as they are put in
the store-room. If they are weighed again the
third day after the birth of the worms, the quantity of those not impregnated will be ascertained.
If the season is hot, you will observe that many
silk-worms are produced in the first 10 or 15 days,
counting from the day of laying the eggs. In
some years I have seen many hatched in this short
space, and sometimes I have observed that these
eggs belonged almost all to the same female.
This precocity is of no inconvenience; it depends
on the particular conformation of the embryo or
of the shell. The egg from which the worm is
come forth is to be distinguished by its white
colour, and because it remains attached to the
cloth.

Upon the cloths where the eggs are, much excrementitious matter will be found, deposited by
the moths. These impurities are not hurtful to
the eggs, provided care be taken not to raise
the cloths before they are perfectly dry.

The form of the cloths on which the eggs have been received is very convenient for keeping them. The fillets of cloth which are raised from the tressels are to be folded into eight doubles, which should form about a foot in width.

These cloths, thus folded, are to be placed in fresh and tolerably dry places, the temperature of which, in the summer, does not much exceed 66°, and does not descend below 32° in winter.

If you are afraid that it should freeze in the place where the eggs have been placed, a thermometer must be put there, or a little water in a dish. If the water does not freeze, the cloths may be safely left there till the following month of March.

During the hot season, the cloths must be occasionally looked at every 10 or 13 days. Sometimes it happens, that when the eggs are too much heaped up in one part of the cloth, and much dirt is mixed up with them, a kind of fermentation takes place, which engenders insects that spoil and devour the eggs. This is to be remedied, and they must be folded up again. I have only once found two of these insects in one of the cloths.

To preserve the cloths always in fresh air, they must be placed on a frame of cord (Fig. 29.), which is attached to the vault or ceiling of a fresh and dry place. In this way the cloths have air on all sides ; the mice cannot get at them, and they are well preserved. They must be inspected nearly every month.

The eggs spoil in a moist place, and the silk-worms which they produce are not vigorous. (Chap. XII.)

When entire broods have been lost, and the origin of the mischief has been examined, it has been easily found that the eggs had been kept in a damp place, which had not been supposed could have been the cause of the loss.

If it is imagined that the place in which the eggs are stored is not dry, it may be ascertained by using the barometer.

Chapter XI.

OBSERVATIONS ON THE VARIETIES IN SILK-WORMS, AND ON THE ESSENTIAL DIFFERENCE EXISTING BETWEEN THE LEAF OF THE GRAFTED MULBERRY TREE AND THAT OF THE WILD MULBERRY GIVEN TO SILK-WORMS OF THE SAME QUALITY.

I HAVE already stated (Chap. III.), that besides the saccharine substance of the leaf which nourishes the silk-worm, these insects also extract, by means of their peculiar organization, a resinous substance, which is purified and gradually circulated into the silk reservoirs, and afterwards spun into silken threads, with which the insects form the cocoon. The insect is therefore, under this view, only to be considered as an instrument for extracting the silky substance from the leaf of the mulberry tree. And they can only extract as much as that leaf will yield.

This being the case, it might be concluded that all varieties of silk-worms are equally good, and it would be unnecessary to examine the advantage or perhaps the loss that might arise from rearing any particular sort of silk-worm.

However, as the term of life of the various species is not alike, and as the different worms afford silk of differing value, it is necessary to explain these differences, the result of which may be of some importance.

There is besides a considerable variety in the quantity of resinous matter afforded by the leaf of the grafted mulberry, and that of the wild mulberry. To state this clearly, I shall speak :

1st. Of the small silk-worm of three casts, or moultings.

2nd. of the large silk-worm of four casts.

3rd. Of the common white silk-worm of four casts.

4th. Of the common yellowish silk-worm of four casts.

5th. Of the comparison between the grafted and wild mulberry-leaves.

1. *Of the Small Silk-worm of Three Casts.*

I reared a certain quantity of these silk-worms separately, the eggs of which may be found in several parts of Lombardy, and close to my own habitation.

These eggs of this species weigh one-eleventh less than the eggs of the common silk-worm ; 39,168 of the latter forming an ounce, while 42,620 of the smaller are required to make that weight. The silk-worms and cocoons of three casts are two-fifths smaller than those of the common sort. My experiments demonstrated that those worms consume, to form a pound of cocoons, nearly as great a quantity of leaves as that eaten by the larger species ; and although smaller when they have reached their highest growth, they devour more fragments and shoots of leaves than the common sort, so that in cultivating the common sort, we perhaps save rather less of the shoots and leaves.

The cocoons of the small silk-worm are composed of finer and more beautiful silk than the common cocoon (Chap. XIV.); but, however, they do not sell higher than the latter. It would appear that the silk-drawing tubes are finer in these silk-worms.

The cocoons are also better constructed, and to this is owing the greater quantity of silk, which at equal weight is drawn from these cocoons, greater in proportion than that afforded by the common cocoons. (Chap. XIV.)

All I have stated should tend to shew that this variety of the silk-worm should be infinitely more cultivated than it hitherto has been ; and those who spin silk, knowing the superior quality of this silk, should give a higher price for it.   Thus would trade benefit ; the industry of the cultivators would be encouraged, who are commonly slow to admit of innovation, and to use discoveries not generally adopted.   Besides these advantages, there exist some equally important.

1st. These silk-worms require four days less of care than the common silk-worm.

2nd. Therefore, the mulberry-trees, by being stripped sooner, shoot again faster, and resist the cold weather better.

3rd. They afford a saving in time, labour, and money.

4th. They are not so long exposed to accidents or contingencies, their life being shorter.

Some imagine the species to be delicate; it appeared to me to be strong and vigorous, but of course requiring care and attention, without which no species will thrive.

As 600 cocoons of this species weigh a pound and a half, and that 360 of the common cocoon make the same weight, it is thought that the worms that produce the 600 cocoons eat more than the common species ; but my experiments shew this to be erroneous.

2. *Of the large Silk-worm of Four Casts.*

I reared many of these silk-worms of a very large quality also separately. The eggs came from Frioul : although these eggs produce larger worms and larger cocoons than the common species, yet are they not either much larger or much heavier ; they are only one-fiftieth more in weight ; 37,440 eggs of Frioul weigh an ounce, whilst 39,168 eggs of the common sort make an ounce.

The worms proceeding from the Frioul eggs, weigh, when at their utmost size, nearly twice and a half as much as the common worm. The cocoons are in the same proportion ; 150 of the large sort weigh a pound and a half, while it requires 360 of the common cocoon to weigh as much.

The only advantage they may offer, is that 18 ¾ lbs. of leaves will produce 1½ lb. of cocoons, whilst 20 ¾ lbs. of leaves are required, to produce

M

1½ lbs. of the common cocoon. And even this advantage is even less in the climate of Lombardy because,

1. In the first place, the silk of these cocoons is coarser and not so pure, (Chap. XlV.), which explains the reason why these worms consume fewer leaves.

2. These worms are five or six days later, in attaining their utmost growth, and in rising, than the common silk-worm.

3. The cultivator runs the chance of stripping the mulberry-trees later, and injuring them.

4. The labourers must be kept on longer, which incurs expense.

5. The insects are exposed to more risk and danger, as their life is longer.

Consequently, this variety in the species does not suit the climate and regions similar to those in which I live; but they may possibly answer better in warmer climates and in different circumstances.

3. *Of the Worms that produce White Silk.*

I have reared a large quantity of this species, separately, and found them in all things equal to the common silk-worms of four casts.

But the white cocoons they produce ought to sell much higher than any others, because it is, undeniably, more valuable than the yellow silk.

The whitest cocoons should be carefully selected for the production of eggs, that the seed may not degenerate.

The art of raising cocoons, being generally totally separate from the trade of spinning the silk, and in different hands, there exists no union between these two classes, which is greatly injurious to both, and retards general improvement: this is the reason so few persons will take the trouble of raising the silk-worms of three casts, or the white cocoons, although the quality of either is superior to that of the ordinary silk.

The white cocoons, far from obtaining a better price, are rather rejected, under an idea that the worms they produce are more delicate than the others, which is entirely an error.

The silk-worms that spin white silk deserve the observation and attention of the cultivator. If I reared silk-worms for the purpose of myself spinning the silk, I would cultivate only the silk-worm of three casts, and the silk-worms that spin white silk as preferable to all others, and every year would choose the very whitest and finest cocoons for seed, to prevent the degeneration of the species.

4. *The common Silk-Worm of Four Casts.*

This species is the most generally cultivated, and that of which we principally treat in this book.

The best are considered to be those that form a pale straw-colour cocoon, in preference to those that make it of deep yellow.

To obtain a pound and a half of these cocoons, it will require 20¾lbs. of leaves of the mulberry, (Chap. XIV.), as we shall shortly show.

The cultivator prefers this species, for the convenience of always having his own eggs, from mere motives of habit, and it is so generally adopted, that I can have nothing further to say on the subject.

5. *Comparison between the Leaf of the Grafted Mulberry, and that of the Wild Mulberry given to Silk-Worms of equal quality.*

I have fed, though with difficulty, a number of silk-worms on the leaf of the wild mulberry-tree alone. This leaf is scarce, because even the hedge mulberry-trees are grafted.

The cultivator finding that the grafted mulberry yields more leaves, is always anxious to graft as soon as possible ; which has prevented my making the experiment on a large scale. I have however, ascertained the following facts :

1. That 14½lbs. of wild mulberry-leaves, weighed when just gathered without sorting, will produce a pound and a half of cocoons ; while, as I above stated, it requires 20¾lbs. of the leaves of the

grafted mulberry, to yield the same quantity.
(Chap. XIV.)

2. That 7½lbs. of cocoons, proceeding from
silk-worms fed on the leaves of wild mulberry,
give about fourteen ounces of exceedingly fine
silk; whilst generally the same weight of silk-
worms, managed exactly in a similar manner, but
fed with leaves of the grafted mulberry, only
yields eleven or twelve ounces of silk.

3. That the silk-worms fed on the wild leaves
are always brisker and have better appetites.

These facts then prove, that the leaf of the
wild mulberry, compared to the grafted mulberry,
yields, at equal weight, a greater portion of nutritious and resinous substance, and less of the
fibrous substance.

I said just now, that I spoke of the leaves,
that they were just gathered and not sorted; because the total weight that is drawn from the
tree must be reckoned, as it is purchased in the
gross, and not sorted until after the weighing and
purchase. (Chap. XIV.)

The fruit of the wild mulberry weighs less than
the grafted fruit, particularly if the grafted tree
is old and the leaf ripe.

Dividing a quantity of leaves of the grafted
mulberry, which were sold as being 150lbs. weight,
into 100 parts, I myself separated 28 portions of

mulberries, 32 of stalks, and 40 of pure leaves.
Thus it is, that this great weight of the grafted
mulberry-leaves, drawn from the trees when the
season is advanced, diminishes considerably, when
we understand these details; and this also accounts
for the large proportion of litter that is removed from the wickers in the fifth age, compared to that laid on the wickers, and compared to that produced in the preceding ages.
(Chap. XIV.)

The result of this must be, that if taking two
trees of equal age and vigour, the grafted tree
yields 50lbs. of leaves, and the wild tree only
30lbs.; on calculation it will appear, that the
weight of nutritious substance eaten by the silk-
worm, will be nearly equal in the two varieties of
mulberry: the silk-worms having the advantage
of being nourished upon a better leaf in that of
the wild mulberry-tree, and which will produce
more silk.

It would appear from what I have stated, that
I prefer the culture of the wild mulberry; however,
before decision, many proprietors should weigh
the following observations duly.

1. In the vast class of the wild mulberry-plant,
there are varieties of an indifferent quality, which
yield few leaves, those leaves much indented and
the branches thorny.   (See the Note p. 31.)

2. Some there are which give a great quantity of fine sound leaves, which can hardly be distinguished from the grafted mulberry-leaf.

3. The *wild* mulberry-tree of a good sound-leaved quality, may be grafted on the *wild* mulberry-tree of the inferior sort with indented leaves.

4. As it is the nature of the wild mulberry to have a great quantity of small branches, that thicken the tree too much, they should be well pruned and thinned, which will also strengthen the tree.

5. The hedge rows of wild mulberry-trees should all be grafted with the best sorts of the wild mulberry, and they should be planted in every spot where they are not likely to injure any other production.

If it is required to increase the production of cocoons, it is indispensably necessary to make every effort to multiply the production of the mulberry-leaf, either grafted or wild ; and in fol lowing the plan I have given, great advantages must inevitably result.

Many cultivators rear their silk-worms until the third moulting, or casting, and sometimes until the fourth moulting, with the leaves of the wild hedge-row mulberry, and the insects eat it with much more eagerness, than that of the grafted mulberry, and it imparts a much greater fragrance to the laboratory.

The hedge rows of grafted mulberries, however, yield a greater abundance of leaves, than is produced by the wild mulberry hedge-row.

The quantity of cocoons must depend on the quantity of leaves. We shall soon show that we may calculate upon obtaining 15lbs. of cocoons from 202lbs. of grafted mulberry leaves. (Chap. XV.) The cultivator must therefore occasionally strive to augment the plantations of the mulberry-trees, either in standards or in hedge-rows, or in short, in every way that does not interfere with the cultivation of his land.

About twenty years ago, in various districts was introduced a method of planting large tracts of land with mulberry trees, at small distances from each other, so as to form wood or coppice, cutting them down every two years, to form full thick brush-wood. To improve these, the soil must be well manured, and moved frequently. I have not personally tried this mode of cultivation, therefore cannot state the advantages it may possess beyond that of the common culture.

Many say this system is delusive, that it can only be recommended where land is scarce, that it only tends, instead of producing fine and stately trees, to load the earth with deformed and unseemly bushes, yielding few leaves, &c.

If in all I have stated, there should be any error, as often happens on other points of agriculture, the cultivators must be accountable for it alone, as they seldom apply the certainty of exact calculation to the innovations they adopt.

As to the hedges of low mulberry-trees formed in indifferent and unproductive soils if cultivated with intelligence, they are of real benefit.

I must repeat here, that which I said and pub lished in Dalmatia nine years ago:—

" You will have a quantity of leaves if you plant " mulberry-trees at given distances, as inclosures " on each side the high roads, in the hollows ; " if you plant hedges of mulberry trees every- " where, not injuring your other fields or produc- " tions, you will have a still greater abundance of " leaves. And obtaining this large quantity of " leaves, there will soon be a larger quantity of " cocoons."

I may have enlarged too much on this subject, and I must conclude, that before the culture of the wild mulberry-tree supersede the grafted tree, experiments must be made for many years, from which alone exact calculations and unerring practical comparisons can be deduced.

This appears to me so important an object, that I shall not suspend the experiments I make every year to attain exact informatiom upon it*.

* The author most obligingly allowed me to visit the large establishment at Varesè for his laboratory of silk-worms, where he has also planted a number of mulberry-trees.
From

Chapter XII.

OF THE DISEASES INCIDENT TO THE SILK-WORMS IN THEIR DIFFERENT AGES, OF THE CAUSES PRODUCING THESE, AND OF THE MEANS OF PREVENTING THEM.

IT was in conformity to the wants and intelligence of man, to create a medical system for himself, to apply unto himself its principles and its remedies. It was also natural that he should create another system applicable to the animals domesticated under him, to which he is indebted for much of his well-being.

The silk-worm being a robust animal by nature, the simplicity of its organization, although it lasts but few days, and managed by the care of man, it would appear impossible that there should have been written hundreds of works upon the subject of its diseases.

From what I observed, his experiments, conducted with great acuteness, upon the cultivation of the mulberry, will in a few years offer the most satisfactory results; if, as he led me to hope, he means to publish a work on the subject, I shall offer a translation of it.

I have thought it best to leave this note which appeared in the first edition, although death having arrested the honourable and useful career of this estimable man has prevented the completion of that work which would have added to the benefits he had already bestowed on the public.—*French Translator.*

If we would explain why so much has been written upon this matter, we should find strong evidence that it is because the disorders have been looked upon as constitutional, and that it has not been considered that they arose from the ill-management and rearing they have undergone.

The silk-worms being reduced in our climates to a domestic state, we should seek to follow nature in our management of them, as nearly as may be practicable, and thus reap all the advantages the cultivation of this insect may offer us. We may then be almost certain of never seeing them attacked by disease during the thirty-five days they require before they attain the period when they pour out their silk; this precious produce, which is one of the chief springs of wealth to our country.

All I have written in the preceding pages of this work should suffice to preserve the insect from all the disorders incident to them.

This consideration made we waver whether I should treat of the diseases of silk-worms, particularly as I had never witnessed any in my establishments, and that I was obliged to seek these disorders in other establishments, managed in the old system, when I wished to observe them. I have, however, determined to write this chapter, particularly to demonstrate the truth and usefulness of the system I recommend. I must here repeat, that whenever it is strictly followed up, the silk-worms will never suffer disease; and that, on the contrary, whenever it is departed from, they will be exposed to the attacks of those disorders I am going to describe.

1. *Diseases which arise from some Defect in the Eggs, or some Error in the Preservation of them.*

1. When the apartment destined for the coming forth, coupling, and laying of the eggs of the moth is too cold, the impregnating liquid will not be perfected and developed only in scanty proportions, at a temperature of only 54° or 59°, and consequently does not sufficiently act upon the eggs, to give them the ashen colour which in the course of 15 or 20 days alone indicates the perfect impregnation. The unimpregnated eggs produce no worms, and those imperfectly impregnated, bear in them the seed of diseases that destroy the silk-worm in various stages of its existence.

2. When the temperature of the said apartment is too hot (77°, 81°.) In this temperature, if the male delays coupling, it looses much of the impregnating liquid. If united to the female too soon, upon issuing from the cocoon, she has not time to evacuate a superabundance of liquid matter, with which she is loaded. She is therefore

disordered, and the impregnating liquid of the male is weakened by admixture with this abundance of liquid matter in the female; consequently the impregnation is imperfect, which produces the same effect as when they are unimpregnated.

3. When the space where the eggs are hatched is too damp, they cannot dry sufficiently, the evaporation of moisture not being free.

The stagnation of this dampness more or less affects the embryo, and engenders diseases analogous to those I have already mentioned.

4. When the place where the eggs were kept and preserved has been damp, the embryo will suffer being kept in a medium which does not allow of the slow and gentle evaporation of the matter contained in the shell, by which it insensibly attains the state assigned to it by nature.

5. When the eggs are too thickly heaped together, in which case, although the place be dry, the transpiration of the eggs will be intercepted, as also the even contact of the air, the eggs heat, and are affected even at a low temperature; consequently the embryo may be injured by any of these causes.

No disease will occur,—

1. If the temperature of the place where the moths are kept be maintained between 68° and 75°.

2. When the apartments are dry.

3. When care is taken to keep the eggs in the cloths on which they were deposited in the proportion of one ounce upon a space three feet square of surface.

4. When the cloths on which are the eggs are not folded too often, not above six or eight times double, and are hung on the frames which I described.

2. *Diseases which attack Silk-worms when the Rules I have prescribed for hatching the Eggs have not been strictly adhered to, although the Eggs were good and well preserved.*

These diseases, numerous and fatal, occur,—

1. When the embryo, just verging toward its transformation into the worm state, in a moderate temperature, is suddenly exposed to a much higher temperature. Its developement is forced, its organs decomposed, and the cocoon of the worm, instead of a deep chestnut, appears more or less red, which is a certain sign of alteration and future disease.

2. When on the point of transformation into the worm, the embryo is suddenly exposed to a lower temperature. The damage is then proportioned to the length of time the exposure acted upon the embryo ; it is extreme if it has lasted many hours.

3. When the silk-worms, being just hatched,

are exposed to a higher temperature than that in which they came forth. The surface of these insects is extensive in comparison to their weight, as the surface of six barrels, of one cwt. each, will be greater than that presented by one barrel, of six cwt. The strong evaporation promoted by heat affects their delicate organs too strongly, particularly before they have eaten.

4. When, on the contrary, the worms, being just hatched, are exposed for any length of time to a much colder temperature than that in which they came forth. Were it only for a few hours the evil would not be of much consequence, but if it last a day or more, the insects get weaker, feed little, and have great difficulty in recovering.

Those persons who hatch eggs for others, and have not warm places in which to put the silk-worms in as fast as they are hatched, run a great chance, if the spring is unfavourable, of injuring whole broods of silk-worms, as I witnessed in the cold spring of 1814.

I observe that the silk reservoirs are the first organs affected by the diseases proceeding from the causes I have been describing. If they are deeply contaminated, the silk-worm will always be infirm, and die before the completion of its natural term of existence.

The alterations and diseases of which I have spoken, cannot occur if,—

1. The eggs in the stove-room were at first in a temperature of 64° to 66°, which may have been gradually raised about two degrees every day, until the complete hatching be accomplished. (Chap. IV. § 4.)

2. If the insects were kept in an even temperature of about 75°.

3. If, in removing the silk-worms, they have been protected from cold air and draughts of wind, particularly from dry cold winds.

3. *Of the Diseases incident to the Silk-worm in its four first Ages, arising from ill Management.*

In this paragraph we must not admit the existence of the seeds of the diseases proceeding from the causes specified in the preceding pages, because, in the contrary case, it must be evident, that whatever might be the care bestowed on the rearing of the silk-worm during the course of its life, we should always find disorder appearing, which would extend through the different ages of the silk-worms, and would destroy them, without it being in the power of the cultivator to remedy the evil.

If we suppose that all rules prescribed have been attended to, the worms can never experience disease.

But in the ordinary manner of rearing silk-worms, the diseases of the silk-worms occur,—

1. When they lay so thick on the wickers that they cannot feed with ease, when they are inclined to do so,—as for example, if on a space which might contain 10,000 worms, 10,000 more are placed, it must be evident, that in this case many insects will feed ill, or not at all; a difference in their developement will result, and large healthy worms will be found mixed with small and sickly worms. This difference, which becomes more palpable as the cause of it is prolonged, engenders disease, and produces death in a great number.

2. When the custom of keeping the worms too thick is in any degree general in an establishment, the inequality of size in worms is not the only ill effect: it will affect the period of their transition, some silk-worms will be found roused, some torpid, and others still requiring food, previous to their transition.

This confusion kills great numbers, even of the strongest; I call those the strongest in the different ages, and particularly in the two first ages, that have eaten most, and have soonest sunk into torpor. To feed those that still require food, and are not torpid, leaves are heaped on the litter, which is already damp. The torpid worms are buried between the old and the fresh litter, and remain in a state of moisture, and covered with excrement.

This condition affects their organs, particularly if the litter is hot and damp. The degree of alteration in these insects may vary considerably, according to the magnitude of the evil causes. A large portion of the silk-worms may rot, or cease to feed, and die in the course of a few days; others continue to eat a little, but shew signs of languor, and dwindle; others may recover to a certain degree, and drag on a sickly existence, without dying. I have mentioned that the first organs injured and deteriorated, even in the first ages of the silk-worm, are invariably the silk vessels. When deeply attacked, the constitution of the worm is quite changed, the secretions are impeded. The insect ceases to be a silk-worm, but is a degraded animal, that can no longer attain the object for which nature designed it.

3. When the air of the laboratory is not renewed, and that the damp stagnates in it, two great evils ensue. The first is, that the transpiration is checked; the second is, that the litter ferments, which increases the heat, the dampness, and corrupts the air, so that the insects become weak, and decay. These evils are aggravated, if the exterior air is mild, still, and damp.

4. If, in addition to the above causes, a wet season should occur, and the leaves be very wet, and they should be given to the silk-worms before they have been sufficiently dried, it will, in all the above ages, quicken the fermentation of the litter, and augment the dampness of the laboratory. In this state, should there not happen to blow any drying north winds, to expel the moisture, the constitution of the silk-worm would, in a short time, be affected, as I observed the case to occur in 1814, a year in which the worms of several laboratories almost all perished.

These losses never can happen,—

1. When the worms are evenly distributed in space proportioned to their numbers, as I have frequently repeated in this work. (Chap. XIV.)

2. By renewing the air, and keeping the laboratories dry, by the means I have indicated.

3. And if care be taken to gather the leaves some time before they are wanted, and when they are wet, to have them thoroughly dried before they are given to the silk-worms.

I might have added, as one of the causes of disease in the first age, and even in the second age, when the leaves are blighted or yellow in consequence of the badness of the season; but this is a case of rare occurrence, particularly if care is taken not to allow the worms to come forth until the season seems favourable, and the shoot of the mulberry-tree is thoroughly developed. If, after all this care, the season should turn out very indifferent, and the leaves bad, the temperature of the laboratory may be gradually lowered for two or three days to 68° or 66°, that the hunger of the worms may slacken; they eat less, and their first age is prolonged some days, which may allow of beneficial alterations in their favour.

It is also, in these cases, equally useful to allow the leaf to wither, that there may be a small portion of watery substance in it.

In the year 1814 many persons experienced these difficulties, and not having used the precautions I have recommended, lost an immense quantity of worms. My laboratories, on the contrary, produced as much as they did any other year, because I never ceased employing the necessary means for their preservation. The second Table, placed at the end of this volume, will exactly shew how I reared the silk-worms, under such circumstances.

It is very rare to experience so very bad a season as that of 1814, even in the cold and variable climates which I inhabit. In warm countries, these astonishing variations are unknown.

4. *Diseases which ill Management will also occasion, in the fifth Age of the Silk-worm.*

In the fifth age, the silk-worms are most subject
to very serious disorders, perfectly unknown in
the preceding ages. The insects being then far
advanced and large, the cultivator must more
especially regret to see them perish, as the
losses are heavier. We shall soon shew that in
these cases also we must attribute the evils to
ill management, so generally practised.

To demonstrate this truth, I am obliged to
enter into particulars upon some subjects that
may appear beyond the comprehension of the
mere practical cultivators. I will, however,
strive to explain myself with such clearness, as
may instruct and convince them. Should I fail
in making myself understood, in treating of the
origin of these diseases, it will at least be sufficient, that I should be comprehended when I
describe the practical means I shall point out
as necessary to prevent disease occurring.

The silk-worm devours, in proportion to the
weight it acquires, an enormous quantity of raw
vegetable substance, compared to that consumed
by any other domestic animal. (Chap. XIV.)

Every animal that feeds upon fresh vegetable
substances, necessarily receives a large portion of
water into its body, as well as alkaline, acid,

and earthy substances, and others, that may be
contained in various proportions in all vegetables.
These substances are generally extraneous to the
necessities of its economy ; for which reason, had
nature not provided the animal with the means
of daily expelling, the oppressed animal would
decay and perish.

Nature furnishes three resources for this purpose,—cutaneous transpiration, pulmonary transpiration, and urinary evacuation.   I do not calculate here the solid and excremental substances
which are voided by the intestinal tube.

It often occurs that one of these means supplies
the others, as in the frequent cases of men who
perspire much and lose little by urinary evacuations.

It must also be noted, that transpiration cannot
exist without the contact of air, and that there is
great analogy between the constituent principle
of urine and transpiration.

The health of animals requires that they should
expel, by means of the excreting organs, the superabundant liquid and extraneous substances which
may have been introduced into their organization
by nutrition.

If the excretions are impeded, the animal will
be attacked by dangerous diseases, such as we see
occur in men from the same general cause.

The silk-worms growing in proportion to the

weight they acquire, devour a quantity of vegetable substance, which, as we remarked, abounds
in liquid, and in substances of which the worm
must disencumber itself.

This is what happens :—This insect has properly neither lungs nor urinary organs. The only
remaining means left it, besides the intestinal
tube, is therefore cutaneous transpiration. It can
well evacuate by this means the liquid, and alkaline, and acid substances which are in a dissolved
and suspended state ; but not having the faculty
of discharging by urine possessed by herbivorous
domestic animals, there remains in the body of the
silk-worm a portion of the earthy particles taken
in with its food, which insensibly accumulate ; of
which we have the proof in the acid and alkaline
earthy substances which the animal evacuates
when it has attained the moth state. From this
there results, that when by want of care the transpiration of the insect is checked, there are certain
chemical attractions formed within it, as yet but
imperfectly understood. It is to these attractions
that should be attributed the various diseases of
the fifth age, commonly denominated in Italian
segno, calcinaccio, and *negrone,* and other similar disorders, which are produced by modifications
of the same general causes that I have before
stated.

The disorder called the *segno* results from the

highest degree of chemical attraction that the silk-worm can endure. It is equivalent to a petechial or spotted disease, and evidently conduces to the decomposition of the primitive animal, and to the compounding of another of a nature totally dissimilar; for when the acid, alkaline, and earthy substances have accumulated in great abundance, and have approximated so far as to exert that affinity termed by the chemists reciprocal affinity, the organic substance is speedily decomposed and disorganized. There are clear proofs of their disorganization in the black spots or other spots which appear on the body of the insect, and are indicative of its approaching transformation into a solid chemical compound; after which the silk-worm hardens and dies. This disease, or decomposition, is never contagious*. As the check of

* In the volume published by M. Dandolo in 1816, in confirmation of his new mode of rearing silk-worms, he gives more detailed explanations of the causes of the disease called segno, which appears to be what in France is called the scarlet (les rouges). The following is an extract from it:—

" It is known that the silk-worm, like all other animals, cannot exist without vital air (oxygen gas), which forms the fifth part of atmospherical air. It is also known that all substances, when in fermentation, discharge a quantity of fixed or mephitic air (carbonic acid); and wherever this acid penetrates, it expels vital air.

" This fixed air, which cannot be inhaled, is, as I said, an acid that may preserve animal substances from corruption, when they possess some principle that combines with this transpiration, and accumulation of the above-mentioned substances, take place more commonly

acid. This phenomenon may occur in some circumstances incident to the silk-worm, in which chemical analysis may discover acid particles, earthy and alkaline substances, that only need an agent, such as an acid, to produce new attractions, from which proceeds the disease termed *calcination*, which is no other than the segno, or spotted disease, in which the chemical action is more prevalent.

" The quantity of carbonic acid which is discharged by the silk-worms is in greater proportion when the worms are numerous, when the temperature is high in the laboratory, and when the atmosphere is laden with moisture. This same acid is well known to be heavier than the atmospheric air ; and consequently it would fill constantly the regions of pure air which are in immediate contact with the silk-worms, if care were not taken to establish currents of air to expel it, which may be proved by using the eudiometer.

" Carbonic acid may for a time strike on the body of the silk-worms without causing death ; because cold-blooded animals do not die immediately on being plunged into this gas, but chemical attractions may ensue very speedily in the interior of the insect, which make it liable to the above-mentioned diseases. If all means are not taken to expel this gas, it continues to act so forcibly, that it may not be possible to arrest the progress of the chemical attractions which successively occur, until at last the animal is in a state of such decomposition, that it becomes a body differing entirely from its natural organization.

" At the period of its existence when it pours its silk, the chemical attractions act most forcibly upon it, and can convert it instantaneously into an incorruptible compound, as is commonly observed."

The disease which the Italians term *giallume*, or jaundice, appears to me to be that which the French call *vaches, gras jaunes;* is caused, according to M. Dandolo, by the action of a certain quantity of moisture mixed with carbonic acid.

N

about the feet, and in the under part of the body, than in the upper part; the former being more re-

The dampness, pressing on the skin of the silk-worm, intercepts transpiration, and consequently impedes the issue of that superabundance of moisture introduced by food into the insects; as, in this instance, the action of the carbonic acid is diminished by the admixture with the moisture, the disease called *segno* does not occur, but the jaundice or *il giallume* will appear.

In works on arts there are repetitions of rules which should not be criticised, and I think the following will not be objectionable :—

To avoid the diseases called *rouge* and *jaunes*, says M. Dandolo:

"1st. Constantly keep a *perpendicular* current of air about the wickers; and should the air in any part of the laboratory appear to want circulation, open the corresponding ventilators. Often make wood fires; and if these heat the laboratory too much, in this case water the floor of the laboratory, which will produce coolness and more circulation of air, because the water will be transformed into vapour by imbibing the heat.

" 2nd. Cool the laboratory as much as possible.

" 3rd. If 'here be any suspicion of disease among the silk-worms, speedily change the litter of the worms, and repeat this change more frequently than is prescribed in the ordinary course of management.

" 4th. Feed the silk-worms on the driest leaves that can be got.

" 5th. Prolong the existence of the worms for some days. Avoid giving them frequent or superabundant meals; because if the food accumulates in the stomachs of the worms in too large a quantity, it may excite too great interior heat, which would facilitate the chemical alteration of the body inclined to disease.

" 6th. In short, in untoward circumstances, it is preferable the worms should be exposed to much air, to cold, to frequent

mote from the contact of the air when the worms
lie too thick, it is also in those parts that the
first symptoms of the disease appear.

Impeded respiration aggravates also the ill state
of the silk-worm.

When the above-mentioned substances occur in
small proportion, and the vital strength is sustained, the worms may have time to draw their

changes of litter, to feed on withered leaves, and even to
hunger. Then will the rouge and jaundice disappear from
the laboratory, or at least much diminish."

More strongly to prove that these illnesses proceed from
the causes to which he has attributed them, M. Dandolo made
the following experiment:—He placed a portion of a bed of
litter, which contained a certain number of healthy worms, on
a dunghill which had almost ceased to ferment, and that
was about 77° of temperature. The air on that day was
still and calm. He examined these worms two days in presence of several of his pupils. The worms were found
spotted with specks of a white saline substance, resembling
the substance of the calcined worms. And, in confirmation
of the assertion, they were agreeably surprised to find some
worms entirely calcined. The experiment was continued;
but there were only found fragments of the white substance
here and there. The worms decaying and rotting, there were
no more that appeared calcined.

This phenomenon was caused, no doubt, by the concurring circumstances necessary for the calcination; for the
worms were not in existence after the first commencement of
the experiment. But the first trial is sufficient (observes
M. Dandolo) to confirm what I asserted, and reason and
science equally demonstrate it clearly. The author earnestly
requests the cultivators of silk-worms to make similar experiments.—*The Translator.*

silk, which is the substance the least susceptible of being affected and corrupted ; but during this operation, or immediately after, heterogeneous substances being no longer in contact with the animal to support or aid it, the vitality diminishes, and the chemical action of these said substances is facilitated.

The worm or chrysalis is then found reduced in a short time to a mummy state.

There is sometimes a species of white efflorescence formed round the mummy, of saline, earthy, and alkaline particles, that have acted upon each other *.

---

* From the moment I knew the disorder called *segno*, and that I particularly observed the calcined worm or chrysalis, I did not hesitate in deciding that it must depend on chemical attractions and combinations, as I before stated. This can hardly be mistaken on seeing the animal texture so altered, and converted into a hard and incorruptible substance.

I removed with extreme ease the white and saline substance which formed the envelope of the calcined worm. I analyzed it ; and besides that, sent it to be analyzed by my friend M. Brugnatelli, Professor of Chemistry at Pavia.

This analysis, as well as that of the earthy substance deposited by the moth when it first issues from the cocoon, were, according to my own opinion, to reveal some very important facts ; and I was not mistaken.

The species of calcination which covers the mummy of the silk-worm or chrysalis in the cocoon itself, is principally composed of the earth termed magnesia, of phosphoric acid, and ammonia, or volatile alkali.

There is no bombic acid found in this composition, which

If the saline particles which envelop the worm or chrysalis can have formed without much moisture, or if the moisture of it have quickly evaporated through the cocoon, the worm will not have spoilt the cocoon, which may be preserved in perfection for a long period. If, on the contrary, these saline particles have retained their moisture, or if much moisture has exuded from the mummy of the silk-worm, and the moisture mixed with the saline substance has remained in contact with the inside of the cocoon, it will soil and stain it, which renders it less proper for spinning, and diminishes its value. In both cases the silk-worm is covered with the white saline envelope.

It however happens sometimes, that the pro-

is peculiar to the healthy chrysalis. It would appear, therefore, that this acid has not been formed, or that it has undergone decomposition, by action of the superior attractions and chemical affinities of the other substances, which afterwards in combination, have formed the saline compound above stated, called by chemists, ammoniaco-magnesian phosphate. This great change in the silk-worm, demonstrating that by ill-management there has accumulated within it a large quantity of heterogeneous matter, against which it has struggled until the end of the fifth age, or beginning of the sixth, presents a strong idea of its prodigious strength of organization; it not only has resisted so great an alteration, but preserved the faculty of pouring the silk it contained, before the chemical affinities could act upon its frame to destroy it, and form a new compound of a substance entirely different from the animal substance.

portions in the heterogeneous substances varying the chemical action does not produce the white saline compound which reduces the worm to the mere mummy state, in which case the cocoon is not injured materially. This alteration is ordinarily called the *Blacks*, (or Negrone.)

There is another disease also called the dark blacks, (or Negrone,) which is produced by another decomposition of the worm. The heterogeneous masses acting on each other, transform the animal substance into a soft compound, resembling soap, and emitting an offensive smell. If the cocoons that contain worms thus affected are quickly spun off, they may be tolerably good, but if delayed, the heat of the weather will engender from this soapy saline matter a number of bristly and disgusting insects, which pierce and spoil the cocoon, and shortly cast their skins and die.

I must here observe, that there are almost always some cocoons that have healthy and perfect chrysalides, mixed with those that are corrupt and affected, and that often on the wicker tables may some worms be found which are attacked by *segno* or scarlet, or by calcination (calcinaccio), when the greatest proportion and number of the insects are strong and well.

The cocoons of which the worm is calcined, as well as those attacked by blacks, (or negrone) weigh perceptibly less than those containing the healthy chrysalis. The chemical alterations to which the diseased worm and chrysalis are exposed, make them lose much weight, as we shall hereafter explain.

The impediment to transpiration, whether general or partial, may occur in divers manners, as well as the accumulation of the deteriorating substances.

1. If the worms lie too thick, the transpiration is checked in those parts which are in contact.

2. If the apartments in which the worms are reared are not sufficiently airy, as fast as the air which is in contact with the body of the worm is charged with damp, it becomes less fit for the insect to breathe, and thus checks transpiration.

3. By the great atmospheric variations. Great heat will promote abundant transpiration; the cold, even when dry, will harden the worm. In the first case, most of the parts of the worm exposed to the air will transpire, while transpiration will almost cease in the parts in immediate contact with other worms. In the other case the hardness caused by cold, instantly checks all transpiration, diminishes the vital strength, and disposes the acid, earthy and alkaline substances to reaction within the insect.

4. In impeding respiration, silk-worms not possessing lungs, breathe, as I before stated (Chap. II.), by several apertures placed near the feet.

When lying too thick on the wickers, they press against each other, and breathe with difficulty; besides which, the transpiration is compressed; the fixed air (carbonic acid) cannot evaporate from the insects, as it does when the insect breathes freely *.

The disadvantages which I have been speaking of, are attributable to ill management, principally in the fifth age, although they may occur in the preceding ages, and cause the animal to become a receptacle of noxious substances, which contribute to breed those diseases of which we have hitherto treated.

Silk-worms are generally reared in a manner so erroneous, that notwithstanding their natural strength, they must sink under such management.

There are other disorders, called giallume riccione, and the soffocamento; the two first are modifications of the above-described, and only differ in violence, or by the circumstances which produced them; however, in these diseases, the silky substance is seldom affected †.

* It would be very beneficial to use wicker hurdles or trays, without ledges or borders. As the carbonic gas, which is heavier than the atmospheric air, might thus escape as fast as it may be formed. The ledges of M. Dandolo's hurdles are not deep, and he no longer doubles the edge of the paper up, but leaves them smooth, to allow free vent for the carbonic gas, through the openings.—*(Translator.)*

† I opened several large sick silk-worms, that would inevitably have died; and I found the silk in the two reservoirs perfect

To prove, for instance, that the disorder of the harpions, or *Riccioni*, proceeds really from derangement in the functions of the skin, curable in some instances, it is sufficient to put several separately, in little unsized paper boxes, that are a certain degree larger than the worm ; and if those boxes are put into a place, the temperature of which is six or eight degrees higher than that prescribed for the particular age, many of these worms will spin a good cocoon, and will

and untouched, without there appearing less of it than in the healthy worm. Having subsequently washed the silk-worms in water, there was deposited a powdery substance similar to that which the moth discharges, on first issuing from the cocoon.

The knowledge of the internal state of the silk-worm in such circumstances, and the analysis of the saline substance enveloping it in the mummy state, termed calcination, were to lead me to important discoveries.

I have collected with care, from the cloths on which the moths were placed, a quantity of this earthy substance which they had discharged, mixed with a liquid substance. It is of a reddish tint, looks very much like mould, and has no peculiar flavour, although it cannot be said to be entirely tasteless. It has a smell very similar to that of the cocoon. I consulted Professor Brugnatelli upon its nature, and the analysis he gave of it presented a most unexpected result. It is a compound of much *uric acid*. This acid is combined with ammonia. Also phosphoric acid combined with lime and magnesia, forming what chemists call phosphate of lime and magnesia: it contains beside, carbonate of lime, and a small portion of animal substance.

The extraordinary part of this analysis, is the uric acid, which had been imagined only to exist in the human *urea*, form a perfect chrysalis. I have, by this treatment, got numbers of good and tolerable cocoons. It was, therefore, evidently necessary, to cure these worms, that they should be sheltered from any agitation of the air, that they should be warmed, and placed in dry air; to conquer the excessive moisture which penetrated them. When damp, heat, corrupt air, fermented litter, and stagnant air, act at once on the silk-worms, it is evident that suffocation may at any time take place, and speedily annihilate their vitality.

from which it derives its name, in the urine of man, which contains *urea,* and in urinary calculi.

Having discovered this acid in the moths of the silk-worm, it might be found in other insects. The solution of the following problem might also be found. How is it that a quantity of this peculiar acid is found in the *guano ;* an earthy substance which exists in various places, and on some of the various coasts of Peru, and which has been used as manure for a great series of time to the inhabitants of that country? It would then be found that *guano* is produced by the decay of an enormous quantity of insects, or of substance deposited during a long series of ages, by insects, particularly as it is known that the *guano* is found in a layer on granite, as ancient as the creation of the world. The other substances which compose guano are analogous to those contained in the earthy substance deposited by the moths of the silk-worm.

To return to our subject, it appears that there is always a certain portion of earthy acid and saline substance retained in the body of the silk-worm, even in a healthy state, which when augmented by ill-management must produce in the insect that series of chemical attractions of which we have spoken above.

I can testify from my own experience that there is a more sudden and dangerous effect to be feared from stagnant, damp, and hot air, than from air stagnant but dry, and at a regulated degree of temperature *.

* The experiments I made, proved to me, that silk-worms being animals without red or hot blood, vitiated mephitic air is not so instantly and eminently fatal to them as very damp hot air.

If a silk-worm is introduced into a bottle filled with mephitic air, in which some of the insects litter had been put, and in which a candle would go out, and a bird die, the worm will live 10, 15, or 20 minutes, although at the end of a few minutes it will appear to suffer, but no warm-blooded animal could stand it half the time. If in the course of a few minutes the animal is withdrawn from the mephitic air, it does not appear at all the worse for it, but seems healthy; therefore it must have extracted some vital air from the mephitic air to have breathed at all.

The silk-worm can also inhale the slightest particles of vital air which water may contain, and can live some minutes immersed in it, particularly when small, and although it may appear dead, it will revive when taken out of the water.

If, however, it cannot find any particle of vital air, it directly dies; for if instead of plunging the worm in mephitic air, or in water, you stop up the eighteen breathing tubes or vessels with grease, the animal will expire instantly.

If a healthy silk-worm is put into a vessel full of vital air, but charged with moisture, and at a temperature of 88° or 98°, it will shortly become flabby, will cease to eat, and perish very soon afterwards.

The warm-blooded animal, on the contrary, such as a bird, will live very well at 88° and 98° of temperature, whatever be the moisture, so it has plenty of vital air.

This proves how different is the organization of these two classes of animals. The functions of life are favourably per-

Thus great numbers of diseases will never appear, if,—

1st. The silk-worms are kept thinly spread out on the wickers, that they may breathe and transpire freely.

2nd. The interior air of the laboratory is constantly and evenly maintained at the temperature I have recommended.

3rd. When the air is never allowed to stagnate in the laboratory, and that it be kept in a gentle slow motion.

4th. By constantly burning blazes when the exterior air is moist and stagnant, and the interior evaporation superabundant.

5th. When the laboratory is kept light, light being the most powerful excitement to living nature.

formed in the warm-blooded animals, when there is a sufficiency of vital air. Its organs are not apt to be relaxed so much as to impede nutrition and secretion. While the silk-worm, if the air is hot and damp, let there be ever so much of it, will pine, its skin will slacken, and the muscles soften, contraction will cease; in time, transpiration will be impeded. The indispensable secretions of life, which in this insect are effected by means of contraction, are suspended.

The skin which covers the worm, is of such contracting power, that when cut, it shrinks as if it had been drawn out, and was elastic.

These reflections will exemplify how essential it is to renovate the air, to dispel moisture in the laboratories, which is the greatest evil that can assail the silk-worms.

6th. By never having the litter liable to fermentation longer on the wickers than I have prescribed.

7th. By being careful never to distribute leaves that have not been thoroughly dried.

8th. By using the fumigating apparatus when needful, the vapour of which destroys the most noxious animal emanations. (Chap. VII.)

These things will be sufficient to prevent the occurrence of any of these diseases*.

* The author, convinced by experience that silk-worms never can be attacked by disorders when reared upon the principles he described, has doubtless omitted entering into any symptomatic details of the diseases. He has mentioned the causes which produce them, and the physical and chemical phenomena that result from them, but has hardly delineated the particular character which distinguishes them from each other, and the symptoms by which they may be recognized, an object which appears to me the most interesting to the cultivator: it might be conjectured, by the silence of the author upon some diseases observed among silk-worms in France, that they must have been totally unknown to him.

Besides, the names given to the diseases known in Lombardy, are almost all in an inverse sense to those given in L'Abbe Rozier's cour d'Agricuture, in which the article on silk-worms is an excellent compendium of all that has been written on silk-worms by French authors, for which reason I have extracted the description of the diseases of silk-worms, which I found in that work.

Of the Scarlet.

This disease is so called, from the more or less dark red colour which the skin of the silk-worm assumes when issuing, or just after they have issued, from the egg. The worms attacked by this disorder, seem cramped, stupified, and suffocated, their rings dry up, and they look exactly like mummies, the red colour becomes ashy and white. This disorder does not always kill the worm in the first moulting, nor yet in the second; and sometimes they do not die until after the fourth moulting or casting of the skin, when they have uselessly consumed a quantity of leaves. When they live so long, it becomes more difficult to distinguish, as their colour becomes less dark and remarkable, and they cannot so easily be separated from those that are healthy, and might be mistaken by the most practised eye. They will even attain as far as rising on the edges, and weave cocoons which are good for nothing, which are vulgarly called *cafignons*, from being soft and ill-woven.

Of the Vaches, Gras, or Yellows.

Some authors divide this disease into three classes, but the specific characters which they give of them do not appear to me sufficiently marked to induce me to be of their opinion. It may be that a difference of names in various provinces may have caused this division of one disorder into three classes. And in one province, it may be further stated, a disorder may present circumstances which it will not offer in another province. But still I persist in thinking, that these classes are only modifications of one disease. The following are the true and essential characteristics of this disorder:—

1st. The head of the worm swells. 2nd. The skin is drawn tight over the rings, and shines as with varnish. 3rd. The rings swell. 4th. The circumference of the aperture of the stigmates is of light or deep yellow colour. 5th. And the worm voids a yellow liquid which may be seen on the leaves.

This disease generally appears towards the second cast, or moulting, rarely later, and is scarcely known in the fourth age. M. Constant du Castelet, one of the earliest and best writers on the management of silk-worms, attributes this sickness to an acid and viscous liquid which penetrates the two reservoirs attached to the sides of the insects, and which mixing with the gum of which they are to spin their silk, impedes the progress and consummation of this gum, and causes a general distention in the insect, which makes it stretch its feet: it soon after becomes soft, contracts, and then bursts upon the litter. The acrid humour, issuing from it, will kill any of the worms that touch it, which seems foreseen by the suffering animal, for they avoid the approach of all the worms, and seek the edges of the wicker, as if to die alone. They are equally avoided by the healthy worms, who seek to retire from the pestilence.

The moment any worms appear attacked by this disorder, it is to be feared it may communicate to the whole brood; they must, therefore, be carefully examined, and when there is the least shadow of doubt, remove those suspected of the disorder into the infirmary, where change of air alone may cure them if they are not too far gone. As to those that positively are attacked by this disease, there is no expedient but throwing them away, and burying them in the dunghill, that the poultry may not pick them up, and be poisoned by them.

Of the White, or Tripe Sickness.

M. Rigaud de Lisle, inhabitant of Crest, is, I believe, the first who distinguished this disorder. The worm, being dead, preserves its fresh and healthy appearance; it must be touched, before its death can be ascertained, and it feels like tripe.

Harpions, or Passes.

These vulgar denominations have passed from the southern to the northern provinces traditionally, when the cultivation of silk-worms was first begun. *Harpions* is derived from *claws*, and *passes* from suffering.

This disease is not really different from the scarlet, it is only a modification of the same; it appears in the earliest stage, a few days after the hatching of the worm, when they assume a yellow colour; that of the *passes* is darker yellow We must refer to what we have described of the symptoms of the scarlet. As to the two disorders of harpions and passes, they are occasioned by the same causes which engender the scarlet. The diseased worms are known, 1st, from their yellow tinge; 2nd, their lengthened spare shape, and wrinkled skin; 3rd, from their sharp and stretched feet; 4th, they eat little, languish, and are in a state of atrophy.

When there are very few worms attacked by this disease, they may be nursed in the infirmary; but as I am persuaded they never will answer, it is better to throw them away; and if before the first casting or moulting the whole brood appear to be infected, then I must insist on the whole being thrown away, and a new brood hatched from fresh seed.

Of the Shining, or Glow, or Luisette.

There are in general but few worms attacked by this disorder. It will appear after the moultings, and most commonly after the fourth moulting. It does not proceed from any defect in the process of hatching, as some pretend, but rather from some failing in the coupling or in the laying of the egg. The worms attacked by this disorder feed like the others, and grow in length exactly in the same proportion, but not in thickness. This disease is perceptible by the colour of the worm, which first appears of a clear red, and then changes to dirty white. If attentively observed, it will be seen to drop a sort of viscous humour from the silk drawing tubes, or spinners, and that its body is transparent, which has occasioned it to be called luisette, or *glow-worm*, the luminous insects that shine in the summer night. The moment these glow-worms or luisettes are discovered on the wicker hurdles, they must be thrown away. These worms consume leaves without ever spinning good cocoons. After the fourth moulting, some glowing luisettes may be found inclined to form the cocoon, moving their heads about, and seeking a place to fix this silk. They must not be allowed to waste their strength and silk in useless efforts; and if they have reached their point, they may have a chance given them, by being put into baskets, with small branches to assist them to form the cocoon.

Dragees, or White Comfit.

This is not properly a disease of the silk-worm, because when the cocoon is formed, when this name is given it, a comfit cocoon does not contain a chrysalis, but a white shortened worm, looking like a sugar-plumb, from which it derives its name. If the worm after forming the cocoon has not had strength to transform itself into a chrysalis, it is a proof it has suffered ; but what has been the disease ? Nobody as yet has been able to designate it. Whole broods of cocoon will be affected by it, and produce scarce any thing but these white sugary worms. However, it is not a total loss, as the silk is of as good a quality as that of a healthy cocoon ; the only loss would be in selling them, as they are much lighter, but if spun at home, profit is equal, as the silk is excellent. A comfit cocoon may be known by shaking it ; the dried worm rattles, and gives a sound which is different from that of the chrysalis. In one of the three volumes published by M. Dandolo, since this work, he makes some observations on two diseases that do not appear to be exactly the same as those we have been describing, called in Lombardy *calcinaccio*, which, however, is very like the white comfit worm disease, and *gattine*.

" Calcination," says M. Dandolo, " is not a disease ob-
" served in any other species of worm, not even in the cater-
" pillars, that live in the open air, which evidently proves
" that it proceeds from evil management. This disease is
" the result of certain chemical combinations which may de-
" compose the component substance of the silk-worm at
" any period of its existence. The causes which produce it
" are such, that sometimes it will declare itself rapidly, or
" sometimes it will remain dormant until the moment of
" rising on the hedge, and even when it has formed the cocoon.
" It becomes general in a laboratory, or is partial, according
" as the chemical element that produces it is spread or con-
" fined to peculiar parts ; but it is never contagious. A worm
" having died by calcination, put in contact with a healthy
" worm, will in no degree affect it."

Some opponents of M. Dandolo, particularly M. Decapi-
tani, curate de Vigano (Lombardy,) having published that calcination was a catarrhal affection produced by a sudden suppression of transpiration, the author, in 1818, made the

ten following experiments to try the disorder, and to demonstrate the error of his antagonist by clear evidence of truth?

1st. He placed the silk-worms of one ounce of eggs in the upper story of the establishment, and left them exposed to atmospheric variations, which were great that year, until they rose on the hedges ; many died, but not one of calcination.

2nd. The silk-worms proceeding from one ounce of eggs were put in a small laboratory after the first moulting, and reared in it until the fifth day of the fifth moulting, without the air being renewed at all in the laboratory: no worm was attacked by calcination.

3rd. In the same laboratory a certain number of worms were put into boxes. After the third moulting the air was so corrupt, it contained not above $\frac{7}{100}$ or $\frac{8}{100}$ of oxygen The worms nearly all died, but without presenting the slightest symptom of calcination.

4th. A certain number of worms, after the first moulting, were reared at 55° of heat, and were gradually used to the temperature. They employed double the usual time in accomplishing their moultings. At the fourth period they were only half the weight of the worms raised at the ordinary and higher degrees of heat. Although they appeared as large, their labour was very unequal, and a part of them were attacked by the disease called gattine, which we shall hereafter describe. The cocoons weighed one-third less than those formed in the usual temperature, but there was not a calcined worm in the whole number.

5th. The silk-worms, reared in 64°, consumed 45 days from the hatching, to the fifth moulting or casting of the skin. They shewed little vigour ; two-thirds of them perished, and they would not have risen on the edge, if the temperature had not been raised. The few cocoons they produced were of inferior quality, many of these being small and light. No worm, however, was attacked by calcination.

6th. Worms, reared at 66°, consumed 40 days from the time of hatching, to the accomplishment of the fifth casting. They had no appearance of vigour ; sought warm places ;

many died while rising on the hedge. The cocoons were light. No calcination was to be seen.

7th. He chose sickly worms ; they were exposed to a high temperature, to excite transpiration in them ; and according to the common notion, thus to ease them of the disorder of calcination, which was supposed to have attacked them. The temperature was raised from 88° to 98°. The worms did not transpire.   Those that were really sick died without shewing any signs of calcination. The rest went through their periods of existence with the greatest regularity.

8th. Some silk-worms were kept too close, and laid thick on the wickers ; they almost all died, but shewed no symptom of calcination.

9th. Some silk-worms spontaneously hatched having been reared in a fluctuating state of temperature, now raised, and then lowered, at each moulting many died ; and out of 3900 there remained only 2600 at the period of rising, and many of these were small and sickly. There were only 1200 cocoons of good quality in the number.

Finally, the silk-worms, proceeding from half an ounce of eggs produced by moths supposed to be attacked by calcination, were reared with the ordinary care, and in the prescribed rules ; the whole course of their existence was regular, they were fine and healthy until the rising on the hedge, and produced perfect cocoons.   And M. Dandolo made his pupils observe, when they expressed their surprise at this phenomenon, that the diseases proceeding from the seed given by weak and delicate moths, are not to be feared as irremediably injurious, neither are they so, except when the seed has been carelessly collected, or not well preserved.

The change of nature of the silk-worm, when subject to the *gattina*, (says M. Dandolo,) is a real animal disease, incident to the silk-worm, as it would be to any other living thing, when exposed to bad food, corrupt air or water, ill management, or in any primitive defect in the conformation of the vital organs. By gattina is generally understood a worm that cannot accomplish the prescribed functions of nature. In whatever degree it is affected, in so much will it differ from the healthy silk-worms. At whatever period the disease attacks them, they shew it by uneasiness, they dislike society, some lose their appetite, others eat plentifully, live long, and then go and die off the tray, or on the edge of it, or even die in the midst of the litter, without strength to retreat.

M. Dandolo is of opinion that there are three principal causes to which this disease may be traced.

1. The alteration of the seed, when it has been ill preserved and moved any distance carelessly. 2. If the process of hatching the eggs in the stove-room was irregularly and ill managed. 3. If the worms have been neglected after hatching by being left too long exposed to a cold temperature, or if they have not been carefully attended to while moulting. " There cannot exist diseases," adds M. Dandolo, " when the egg is well impregnated, well preserved, and the silk-worm well attended to. Thousands of experiments have demonstrated this fact."

CHAPTER XIII.

OF THE LOCALITIES AND UTENSILS REQUISITE IN THE ART OF REARING SILK-WORMS SUCCESFULLY.

IT is difficult to imagine how, in the lapse of some centuries, the practice of the useful and precious art of rearing silk-worms should have remained in the hands of the ignorant and the illiterate.

While it is evident that the abundance and certainty of the annual produce of the cocoons rests entirely upon the perfect cultivation of the silk-worm during the various periods of its existence, and as it is generally known that these insects are not natives of our climes, and only exist by the care bestowed on them in their domestic state, it can scarcely be credited that there should not yet be a code of sure rules to form habitations suited to their wants, and favourable to their progress, and that silk-worms are to be exposed to every circumstance most injurious to their health and well-being.

It seems never to have been imagined that four or five ounces of eggs would produce 150,000 or even 200,000 silk-worms, that would all require room to breathe freely of pure air, and to secrete the substances necessary to life.

A building wisely constructed, upon the fixed principles of art, where the air may circulate at all times, and in all cases, and preserve its dryness, would alone powerfully contribute to the constant prosperity of the animal; and in course of time, to the abundant production of cocoons of the finest quality.

When the habitation of the silk-worm has been well prepared, an unspeakable advantage has been obtained, and all will probably advance favourably.

As we must suppose that many proprietors will have laboratories constructed, so as to ensure an abundance of cocoons, I will here give a short detail of the construction of a laboratory, and point out a few indispensable alterations and reforms that may be made in the laboratories of tenants and cultivators which are already built ; and these reforms are infinitely to the advantage of the proprietors and cultivators.

In speaking of these two classes of laboratories, I must also mention one of the most necessary appendages to such establishments ; the place rēquired for preserving the mulberry leaves fresh and good during two or three days, by which means the loss which might occur from being obliged to feed the worms upon wet, or withered, or fermented, leaves, will be entirely avoided.

If the construction of the building which is to be occupied by the silk-worms during their lives is of importance, it may not be useless to give a description of the utensils and tools most likely to facilitate all the necessary operations which are to be executed.

We shall, therefore, in this chapter, speak,—

1st. Of the laboratory of the proprietor.

2d. Of the laboratory of the cultivator, or tenant.

3d. Of the proper place for the preservation of the leaves.

4th. Of the utensils.

1. *Of the Laboratory of the Proprietor.*

It is a certain fact, that men scarcely ever employ capital excepting in the manner most likely to bring in rent, or interest in money, pointed out by habit and local circumstances, with the further exception of the expenses of luxury, &c.

For instance, if a proprietor purchase fifteen or twenty acres of land, which cost him three or four thousand francs, it is with the view of drawing 150 or 200 francs from it.

Admitting this position, it is clearly evident, that if we can prove to the proprietor that in employing capital to the amount of three thousand francs in constructing a laboratory of silk-worms, he would derive more than five per cent., we cannot doubt that he would endeavour to do so.

To be able to say, with assurance, that it is profitable to construct a laboratory for silk-worms, I shall only make one observation, which is, that to draw 150 francs from 3000 francs capital employed in erecting a laboratory, it would be sufficient, that the proprietor received 90 pounds of cocoons more than he annually

obtains in the common manner of managing silk-
worms.

Are there not, besides, many proprietors, who
might arrange, with very trifling expense, such
places as garrets, warehouses, &c., so as to be on
a good plan for the rearing silk-worms?

Can any liberal proprietor, who has informed
himself upon the improvements I have suggested
in the art of rearing silk-worms, state whether
he does not think that from ten ounces of good
well-preserved eggs we may easily draw 150, and
even 300 pounds of cocoons beyond what may be
procured in the bad and ordinary manner of
managing silk-worms.

This interest, derived from such an amount of
capital, is still very trifling, when compared
with the other manifold advantages which the proprietor will derive from the erection of a laboratory. To be convinced of this, it is only necessary to observe,—

1st. That the erection of laboratories tends to
increase the value of the lands in which they are
erected, and from which the leaves are procured,
because they afford a greater produce in cocoons;
as land which, when well-worked and dressed,
gives eight for one, is more valuable than land
which will only yield six.

2d. That in uniting in the same establishment

all the operations relating to silk-worms, there is a vast saving in leaves, in fuel, and in labour.

3rd. That the common cultivators very soon get into the proper manner of managing a laboratory, when they see the favourable results produced by the prescribed rules, and because they find themselves compelled to act in obedience to the directions of the enlightened persons at the head of the establishment.

4th. That as some of the inequalities in the advantages obtained among the common cultivators disappear sooner than others, advantages which are commonly called luck, in rearing silk-worms, having all used the same means to ensure success in the operation, they will adopt those that are invariably secure.

5th. That the proprietor and cultivator, assured of success, will not neglect the culture of the mulberry-tree, nor destroy the tree, as is often done in those parts where the proprietor strips his trees to sell the leaf, in the persuasion that this is more profitable than employing it himself.

6th. That the robbery and dishonesty which may be carried on when the proprietors have silk-worms reared in several different farms, cannot take place so easily.

7th. In short, that the space allotted to them

o

is sooner clear and free of the worms, and may
be made use of for other domestic uses.

What I have said here will demonstrate the
benefit resulting from proprietors erecting large
laboratories*.

My laboratory is constructed on principles
I shall describe, and can contain the produce of
twenty ounces of the eggs of silk-worms, that
is to say, might yield about 20 cwt. of cocoons ;
there is an engraving at the end of the book.

The laboratory is about 30 feet wide, 77 feet
long, about 12 feet high, and, when reckoned to
the top of the roof, 21 feet high. (Tab. 1, Fig. 1.)

Six rows of tables, or wicker trays, of about
two feet six inches in width, may be placed in
the breadth of the room (g). As these wickers
are placed two and two, there only will appear

* There is at this moment, in this part of the country, a
large laboratory, belonging to a proprietor, which is sufficient
to contain the worms of twenty ounces of eggs. It yielded,
in 1813, above 120 pounds of cocoons for every ounce of
eggs. It is situated at Moraggone, and is the property of
an eminent agriculturist, who has there united, in a central
point, the whole labours of his tenants. It required not
any vast expense to erect this laboratory. He has only
made use of a large granary, which he has adapted to the
purpose, and for which he has had all the necessary utensils
made. This place serves as a laboratory during one month,
and then is again used as a granary during the remaining
months of the year.

three rows of wickers, and there will be four
passages or avenues between these three rows,
two next the two side walls, and two between
the wickers. These passages are about three
feet wide; they are useful in giving free access to
the wickers, and for placing the steps and boards.

There should be posts driven in between the
wicker trays, which, as I before said, form an
avenue. On these posts should be fastened little
bars of wood, horizontally placed, which support,
the wicker hurdles; there is, between the two
wicker hurdles, a vacuum of about five inches
and a half, to allow the air to pass freely, and
this space corresponds to the size of the vertical
posts, which form a line at equal distances down
the laboratory.

There are in the building thirteen unglazed
windows, with verandas outside, and with paper
window-frames inside; under each window, near
the floor, ventilators, or square apertures of about
thirteen inches, that they may be closed by a
neatly-fitted sliding pannel, so as to enable you,
at will, to make the air circulate, which, entering
and issuing, will blow over the whole floor.

When the air of the windows is not wanted,
the paper frames may be kept closed. The verandas, or Italian shutters, may be opened or
shut, according to circumstances. When the air
is still, and the temperature of the interior and

exterior are nearly equal, all the window-frames
may be opened, and the Italian shutters must be
closed, at least the greatest part of them.

I made eight ventilators in two lines in the
floor, and in the ceiling, placed perpendicularly
opposite each other, in the centre of the passages
between the wicker hurdles. These ventilators
had sliding pannels, made of thick glass, to close
them and admit light from above, and may also
on some occasions be covered with white linen;
they may be opened or closed, according to circumstances.

As the air of the floor-ventilators ascends, and
that of the ceiling-ventilators descends, according
to the variations of temperature, it must necessarily pass through the three rows of wicker trays.

I have also had six ventilators made in the floor,
besides those under the windows, to communicate with the rooms beneath (d).

All these ventilators should open easily when
wanted, as they can alone maintain a constant
renewal of the exterior air, without the necessity
of ever opening the paper window-frames, which
are within the Italian blinds or shutters.

I here use thirteen windows (2), three of which
are placed at one end of the laboratory, while at
the opposite end there are three doors constructed
so as to admit more or less air as may be required(a).
By these doors we are admitted into another hall

of about 36 feet long, and 30 feet wide; which forms a continuation of the large laboratory, and which also contains wicker tray stands sufficiently raised to facilitate the necessary care of the worms in the laboratory. There are in this hall six windows (2), and six ventilators under the windows nearly on a level with the floor (*b*), as well as four ventilators in the ceiling.

There are six fire-places in the great laboratory, one in each angle, and one on each side of the centre (*h*).

I had a large round stove of about three feet eight inches wide, and nine feet two inches high, placed in the middle of the laboratory (*g*); it divides the large row of the wicker hurdle stands.

I use small glass oil burners or lamps that yield no smoke, to light the laboratory in the night.

The floor of the laboratory is the only one that is covered with Italian cement (ghiarone) or stucco. That in the hall is made of bricks, that in case of necessity the leaves may be dried on it when they are wet with rain; and the eggs may be dried in it after they have been washed in the first process.

Between the hall and the great laboratory, there is a small room having two large doors: the one communicating with the great laboratory (*i*),

the other with the hall (*a*). In the centre of
the floor there is a large square opening, which
communicates with the lower part of the building
(*c*). This is closed with a wooden folding-door,
which may be removed at pleasure ; this aperture is used for throwing down the litter and
rubbish of the laboratory, and is also useful for
admitting the branches and leaves of the mulberry,
which can easily be drawn up with a hand pulley.

This same aperture keeps up the circulation of
air, by moving a large column of air, when the
three window frames at the end of the laboratory
are open.

I have had an interior and an exterior bell hung,
o hasten attendance, when the orders are to be
given or executed.

Such is the construction of my great laboratory, in which I place the silk-worms after the
fourth casting or moulting.

It is impossible the air should remain stagnant
in it, or that it should ever be damp; as this
building stands alone, the ventilator in the three
walls must by their exposition keep the air in the
utmost equilibrium. Should it, notwithstanding
all this, have a tendency to stagnation, the air
may be instantly put in motion by establishing
great currents of air, and burning blazing light
fires in the six fire-places.

When fires are not needed, nor the chimneys wanted as ventilators, they may be closed with chimney-boards.

By the sliding panels on the ventilators that are under the windows, should there be too much draught, they may be closed, so as to regulate the air as it is required. The same is applicable to the high ventilators, with this advantage, that as they can have a glazed window panel, they may be closed and yet give light. Both may be opened occasionally.

The stove is only used when the air of the laboratory wants warming, and the temperature raising.

In which case, when the stove warms the laboratory, a column of external air enters continually into a portion of the body of the stove, which is separated from that part in which the fire burns, and from that through which the smoke issues; in this receptacle, when the air is heated, it escapes, through holes perforated on purpose, into the laboratory, and augments both the quantity of heat and fresh air.

In different parts of the laboratory I have placed four barometers, six thermometers, and two *thermometrographs,* to shew what is to be done in case of any accumulation of moisture,

and of augmentation or diminution of temperature in the laboratory*.

Let us now examine a moderate-sized labora-

---

* Since the publication of this work, there have been
erected large laboratories in Lombardy, which are called
Dandolieres, an honourable testimony to the philanthro-
pical Parmentier of Italy. Some have offered favourable
results; however, some proprietors were disappointed, and
the utility of the laboratories has been called in question.
Divers objections were started, which I shall state, with
the answers which were given by M. Dandolo.

It is said that when a disease appears in the great laboratories, it is propagated with greater facility; and that all
the silk-worms then inevitably perish, which would not occur
were they divided into smaller and separate apartments.

M. Dandolo first remarks that there exist no contagious
disorders among silk-worms, and if all the silk-worms of a
great laboratory die successively, it must be solely owing to
some incipient cause, which killed the first, and which continues to act on them all as long as it remains unremoved.
" Is it not well known that several palpable causes may
strike a number of men, such as bad air, corrupt water,
unwholesome food, and that as long as these causes are not
removed, without being contagious in the least degree,
these vitiated causes will continue to commit general ravages?   A number of facts have long ago proved, and particularly in 1816, that all the silk-worms of a whole province may die, although reared in small separate laboratories.
The mortality in large or small laboratories depends always
and entirely upon the management of the establishment and
the regularity of it.   The only difference will be this: that
an ignorant cultivator will only lose the small number of
silk-worms in a small laboratory; whereas if the director of
a large laboratory be equally ignorant, he will of course lose
twenty, thirty times as many worms, in proportion to the
size of the establishment and his own ignorance.   This is

tory calculated to contain five ounces of eggs only.
The middle-sized laboratory (Table I. Fig. II.,)

the only view that can be taken of the subject. Besides the
proprietor must never forget that large laboratories are only
intended to receive silk-worms that have been well reared
and managed in small laboratories until the third moulting,
or better still, to the fourth moulting." M. Dandolo has no
doubt that the failure of the large establishments has been
caused by the silk-worms transported into the great laboratories not being originally healthy, and having been ill-ma-
naged previously in the small laboratories.

That all be done with order and expedition in the great
laboratories, M. Dandolo divides them; into numbered sections, which are called *Laboratory*, *No. 1*, *No. 2*, *No. 3*,
&c. &c. Each section is composed of a certain number of
tables or wicker tray-stands of 150 or 200 feet square of
surface, and must be attended by an active and intelligent
manager, thoroughly practised in the business, and there
should be a chief manager or overseer at the head of the
great laboratory, who must also overlook the managers of
the sections. On the wall near the door of each section
there should be put up a board with the name of the manager
of the individual section printed on it ; also the names of the
assistants employed in it. A clock should point the hours
for feeding the worms ; and a bell should warn the work-
people when to bring in the leaves and food, and, in short,
mark the usual course of operations. The proprietor or
chief manager should now and then inspect the wickers, to
ascertain whether the leaves have been well or equally distributed, and if all the work has been done well and or-
derly. "Persons who have viewed my great laboratory
(says M. Dandolo), may have convinced themselves that
well understood good management animates the whole system, and preserves from every evil.

The great laboratories or *Dandolieres* present the following advantages:

1. A large laboratory for 20 ounces of eggs, will require
one-third fewer persons to manage it, than will be wanted

which produces about six cwt. of cocoons, should
be 40 feet long, 18 feet wide, and 13 feet high.

to manage the same quantity of eggs when divided in six or
seven small laboratories.

2. There is a smaller quantity of fuel consumed in a
large laboratory than in several small ones.

3. One single man may with greater ease overlook one of
these large laboratories than several separate small laboratories.

4. The large laboratories effect a great saving in the
consumption of the leaves.

5. The circulation of air is better.

6. Whatever be the exposition of the great laboratory,
the interior temperature, in equal circumstances, will be less
susceptible of sudden changes than the interior temperature
of the small laboratories. Neither is it liable to breed so
great a quantity of carbonic acid gas, or mephitic air, so
peculiarly noxious to the silk-worm, as the small laboratory.

7. Finally the silk-worms succeed better in the great laboratory, and form finer cocoons.

The total result will shew that the great laboratory is less
expensive, and better overlooked, consumes less leaf, preserves
the worms better from the attacks of disease, and affords an
abundant and finer produce of cocoons. M. Dandolo observes, that having reared silk-worms in large, middle-
sized, and small laboratories, the cocoons of the large laboratories were invariably of a superior quality. The art of
rearing silk-worms will require some knowledge of physic
and chemistry, and the advantage or success to be obtained
will depend much on the instruction and information of that
person who directs the management, as well as a certain degree of intelligence in the lower workmen. Thus M. Dandolo particularly recommends the choice of persons of good
capacity. The education necessary is not very long, and
the reward is great, as it is the principal and surest income
of the proprietor. These two advantages should surely be
sufficient to induce us to adopt the rules pointed out by the

Six wicker hurdles may be put one above another on each side, and in the length there should be a row of wicker hurdles at about two inches from the wall, to allow the air to circulate round them. These hurdles should be thirty inches wide. Down the centre of the laboratory there are two rows of wickers, of above 33 inches wide each; there is to be a distance of one foot ten inches between the rows. This distance is sufficient to allow free passage to and fro, and as the posts and horizontal boards that support the hurdles form a sort of ladder by which the highest hurdle trays may be reached with convenience, one

author. More quickly to disseminate his improved method of rearing silk-worms, M. Dandolo requested the great proprietors to send him pupils whom he instructed in his laboratory. These pupils often occasioned great losses, as he, to give them practical skill, allowed them to act alone. This occurred particularly in the distribution of leaves in the fifth age, at which period the worms proceeding from 20 ounces of eggs, occupy a surface of 3000 square feet; so it may be easily conceived how serious a loss might be occasioned by the waste of mulberry leaves when it is not distributed with carefulness. M. Dandolo lost many thousand pounds of the leaves by the waste of his pupils during the fifth age. " But it signifies little (this philanthrophist would say), compared to the advantage of generalizing and naturalizing the improved art of rearing silk-worms by means of these pupils." These pupils cost the proprietors but very little during the short season of the management of the silk-worms, and returned, after this trifling apprenticeship, capable of managing a large laboratory and establishment of silk-worms.— *French Translator.*

half of these hurdles may thus be fed and attended to, while those on the other side may be reached by means of portable steps.

There are only two passages down this laboratory, because I do not reckon the narrow separation between the hurdles of only the width of the posts.

There are four ventilators in the ceiling, above the passages, that the exterior air may not strike directly on the hurdles. There are eight ventilators on a level with the floor, placed at regular distances, four fire-places in the four angles, a stove in the middle of the laboratory, the pipes of which may be carried along the walls, and another in the centre of the end wall facing the door.

There are two barometers and four thermometers. In the night it is lighted by two Argand lamps that yield as little smoke as possible.

This laboratory, like the large one, is detached on three sides: there are four large windows, without blinds or shutters, because the whole change of air is made perpendicularly from top to bottom, and bottom to top. It is not necessary there should be ventilators in the floor corresponding to those in the ceiling. However calm the air may be, it may have as much motion as is desired imparted to it, by burning blazes in the fire-places, and opening the ventilators.

I have also small laboratories, that only contain

about 367 feet square of wicker hurdles, and which consequently cannot produce more than 240 pounds of cocoons (Fig 3.); they are only small, thirteen-feet-square rooms, containing in the middle four rows of double wicker hurdles placed one above the other, and about thirty inches wide.

There are four ventilators in the ceiling, two fire-places in the two cross angles, and three ventilators in the walls, on a level with the floor. Each small laboratory has its barometer and two thermòmeters. I have not mentioned the aspect of my laboratories, because all aspects are good, provided the air circulate freely everywhere, and that there be a possibility of shading or shutting out the sun. I have them in all positions.

The cleanliness of my laboratories is extreme; there is no disagreeable smell of any sort, and it requires no perfume. The best is the natural aromatic fragrance of the mulberry leaf while the silk-worms are alive, and afterwards that of the cocoon when forming, or when formed.

In case the season becomes very cold, as it happened in 1813 and 1814, the fire-places may be used, not only to renew the air, but also to warm it, being careful in these instances to burn good large wood, which keeps the fire longer, and not

straw or chips. The stoves are, however, better
for heating the air than the fire-places.

Whichever of these methods may have been
adopted in the choice of the size and construction
of a laboratory, it will be practicable, by the aid
of the barometer and the thermometer, to neutra-
lize or destroy the influence of cold, heat, wind,
stagnant air, damp or corrupt atmosphere, and
the fermentation of the litter even may be prevented or arrested.

2. *Of the Laboratories of Cultivators.*

In general, the laboratories of the tenants,
farmers, and common cultivators, have the appearance of catacombs. I say, in general, for
there are some few, who, although they may not
have all the requisites for rearing silk-worms in
perfection, yet have care sufficient to preserve the
worms from any very severe disease.

I have often found, on entering the rooms in
which these insects were reared, that they were
damp, ill-lighted by lamps fed with stinking oil,
the air corrupt and stagnant to a degree that impeded respiration, disagreeable effluvia disguised
with aromatics, the wickers too close together,
covered with fermenting litter, upon which the
silk-worms were pining. The air was never re-

newed except by the breaches which time had worn in the doors and windows ; and that which made this seem more sad and deplorable, was remarking, that the persons who attended to these insects, however healthy they might have been when they entered on the employment, lost their health, their voices became hollow, their hue pallid, and they had the appearance of valetu- dinarians, as if issuing from the very tombs, or recovering from some dreadful illness.

All buildings are good for rearing silk-worms, provided that, in proportion to their size, there be one or more fire-places, two or more venti- lators in the ceiling, and on a level with the floor, and one or several windows, or apertures, by which light may constantly come in, and yet not sunshine.

My tenants' laboratories, which contain the worms proceeding from four ounces of eggs, have two fire-places placed in the angles, a stove, four ventilators in the ceiling, and three in the wall, on a level with the floor. I have laboratories that only hold the worms proceeding from three ounces of eggs, having two small fire-places, three venti- lators in the ceiling, and two in the basement. I have, besides, laboratories that are calculated for two ounces of eggs, with two small fire-places, two ventilators above, and two in the lower part,

In each of the laboratories there is a large door, in which a smaller one has been made, and two windows, or apertures, to admit light.

When the sun strikes on these, the interior blinds or shutters should be closed.

I must observe, that these slight improvements and alterations in rooms are important, cost little, and produce great benefit.

When the apartment is capable of containing the worms of four ounces of eggs, a brick stove is indispensable; it warms more thoroughly than fire-places, and consumes less fuel. Fire-places, in general, are only used for burning blazes, and wood is only burnt in them when the exterior temperature is very cold indeed, and continues such for any length of time. The fire-places should be boarded.

As soon as the allotted space is arranged as I have described, the cultivator should not be allowed to put the silk-worms in the kitchen, or anywhere but in the laboratory, well spread out on the wicker tray-stands: those who do not insist on this will stand a chance of seeing the greatest part of the silk-worms sick or dead before they are full grown.

These misfortunes, to which cultivators are exposed, arise from their making the silk-worms pass through sudden and violent transitions when they are young, and keeping them too confined and close. (Chap. XII.)

However crowded the worms may be, if the common cultivator sees them move and feed he is satisfied they are doing well ; he does not distinguish those that are squeezed against others, and cannot feed easily, or, indeed, at all, and they sooner or later perish under the litter that covers them.

In each cultivator's laboratory there should be one or two good thermometers left there constantly.

It may not be superfluous to remind the reader, that should any worms be slow in rising, and more backward, they should be removed elsewhere, that equality may be preserved. (Chap. VIII. § 5.)

It would be advantageous to the proprietor to give the cultivator a thermometer and a barometer, teaching how to ascertain the degrees of heat and moisture of the laboratory. The cultivator will then be better able to use those means efficaciously which will preserve the air in the state most favourable to the progress of the silk-worm. And, although the fumigating bottle be not indispensable, yet, as at some periods the animal exhalations leave an unpleasant smell, it may be very useful to the cultivator, the vapour it yields being a desirable corrective.

As to the aspect of the cultivators' laboratories, there is no doubt the coldest is the best, and that which is most exposed to the wind ; and that they are better on the first, than on the ground-floor But, however, if the improvements I have suggested are made, such places as are convenient may be employed with a certainty that the air will never, when my system is followed up, be stagnant, damp, or mephitic.

One might be tempted to imagine that one of the causes that have contributed to the perpetuation of the use of the worst and most inconvenient erections for the habitations of silk-worms is, there having now and then been collected a tolerably abundant crop of cocoons.   It was never thought to be the effect of chance, or, more properly, the effect of the favourable variations of the weather.

One year for instance, in which the season had been fine and dry, in consequence of a north wind, which blew much, it would not be surprising that the worms succeeded, although ill managed.

We have an example of the benefit of constantly dry air, and of air constantly in motion renewed, in the sheds or barracks of the mountainous districts in which they rear silk-worms. They always are more successful than in the plains.

We now find, that let the meteorological influ-

ences be what they may, the fair produce of cocoons can never fail; and that the cultivator
obtains success and profit not by lucky chance,
but by certain and fixed calculation.

3. Of the Building destined to keep the Leaves of the Mulberry-tree fresh and good.

According to my idea, the advantage and damage
has not yet been sufficiently calculated, which
may arise from the leaf of the mulberry, by its
proper preservation or neglect, previous to the
distribution of it to the worms.

It should be laid on the ground-floor, or in
cellars, slightly damp, and which may be closed,
so as only to admit light sufficient to see where
to put it, to stir it about, sort, and pick it thoroughly. These circumstances are indispensable.

1st. Because the lower rooms are always
cooler than the upper floors.

2nd. Because in damp places the leaf is not
exposed to evaporation, which alters and withers
it. I have kept it three days in such places as
I have described ; it diminished but little in
weight, and was not faded. When it is still
very succulent (see note in p. 31,) it should be
laid in layers of two or three inches, that it may
not alter or ferment. When quite ripe it will
keep several days, although the layers be above
a foot deep, provided it has been gathered when

thoroughly dry. It should, however, be carefully moved and stirred about every day, that it
may receive the contact of the air, and not get
pressed down.

If, in the place allotted for placing the worms,
the air ought to be invariably dry in that destined to keep the leaves, it should be cool, damp,
and still. It would be a perceptible loss if the
air robbed the leaf of its natural moisture; not
so much because it would wither it, as that I
consider that natural moisture as a necessary
vehicle for the various separations and secretions
required for the health of the silk-worm, and for
the perfect deposition of the silk in its reservoirs.
Nature has bestowed much less liquid or watery
substance on the mulberry leaf, than on any
other leaf of any tree in our climates.

If the allotted space for the preservation of
the leaf is very damp, it will not alter it, so
that it be cool and well closed ; it is heat and
accumulation that spoil the leaf. It should be
managed that these places should be under, or
very near the laboratories. There will be found
to be great advantage in having a good provi-
sion of leaves at hand, particularly at the voracious period, called in France (*La granda frèze,
ou brife*) in the fifth age of the worm ; and this
advantage will be felt still more, should there
occur any continued rains at that season.

When the proprietor once thoroughly acknow-
ledges the exact and regular method of rearing
silk-worms as the best means of prosperity and
gain, he will find all difficulties cease, and that
the necessary care is attainable, without great expense or much trouble.

4. *Of the Utensils required in the art of Rearing Silk-Worms.*

Better to execute the various operations which
form an art, upon smaller means, ought to be
the principal object at which those who practise
that art should aim.

Upon this principle I have thought it might
not be without usefulness, to form a small collection of utensils which are not expensive, and yet
are indispensable in the execution of the operations which the cultivation of silk-worms requires.

This art has hitherto had no appropriate utensils; each employed whatever came under the
hand, indifferently. I here give an explanation
of the utensils, and the engraving will be found
at the end of the volume.

Utensils.

The Scraper.—It is used to detach, or scrape
the eggs from the wet cloths. It is easily
handled; the sharp side of the blade is introduced

between the egg and the cloth, aslant, and thus a quantity may be scraped in a short time. (Fig. 3.)

The Thermometer.—I have described it and its uses in a separate paragraph. (Chap. IV. § 2.)

The Stove.—It is intended to warm the laboratory. It heats much better when constructed on the principle of receiving external air, heating it, and then dispersing it in the laboratory. The rarefied air when it comes in hot, is a purifier, as it expels the interior air. If required, the holes or apertures through which the rarified air passes may be stopped up, when there is fire in the stove. These apertures may be used as ventilators, to admit cold air, when there is no fire in the stove. (Fig. 5.)

Small Boxes or Trays for Hatching the Eggs of the Silk-worms.—There should be some of all sizes, that each ounce of eggs may have a space of seven inches and four lines square. They should be made of thick pasteboard, if they are small; and if larger, and intended to hold ten or eleven ounces of eggs, of thin board. They should be numbered with very visible ciphers marked on the sides. (Fig. 6.)

Hurdles, or Trays, or Table-stands.—They are used covered with paper, to hold the worms; mine have the surface made of cane. They may

be made of any branches, wicker, or wood, or
basket-work; so that it be not close, but woven
open, to admit of the air being in contact with
the paper underneath, which keeps it dry. The
breadth of hurdles or tray should be from 29 to
37 inches, the length from 18 to 24 feet, and they
should be of equal sizes, that when put above one
another, they may not interfere with the square
baskets, or extend out, so as to be inconvenient.
(Fig. 30.) On the ledge or borders of these trays
may be painted in letters, the number of square
feet contained on the surface; for instance,
supposing the wicker-tray 20 feet long, and 3
feet wide; the numbers to inscribe would be " 30
feet square," &c. (Fig. 7.)

Spoon.—It is made so as to stir the eggs with
care. (Fig. 8.)

Small portable trays, made of thin board, of
about a foot in width, and sufficiently long to
fit across the breadth of the wicker tray or·
hurdle. The handle should be fixed in the centre, so as to allow of their being carried firmly
with one hand. They must be very smoothly
finished, that the silk-worms may get upon them
without difficulty. The ledges round three sides
about half an inch deep. (Fig. 9.)

Ventilators.—Apertures covered with panels,
which should slide up and down easily, by means
of two grooves, in which they are fitted; when

in the lower part of the wall they should draw up, and when in the ceiling slide down. (Fig. 10.)

Piercing Iron.—This tool should be made like a stamp; and by hammering on it, will cut round holes in doubled paper placed on a block. Several sheets of paper may thus be stamped at once, which are used to put over the egg boxes, when the worms are just hatched. Many employ a thin veil, or very coarse muslin; for this purpose, and with the same effect, it is unimportant which is adopted. (Fig. 11. *a*. and *b*.)

Hook.—Small instrument made of bent iron, is very useful to take up neatly and quickly the small branches covered with silk-worms from the boxes, to put them on the sheets of paper prepared to receive them in the small laboratory. With this instrument, it is not necessary to touch the worms with the hand, and they avoid the risk of being crushed. (Fig. 12.)

Travelling Case.—In this pack or box the hatched worms of 20 ounces may be moved to any distance; it weighs about 75 pounds. It is divided into drawers, or shelves, and when there are fewer worms, some of the drawers may be taken out to give more space: each slide or drawer can hold a sheet of paper containing the worms of an ounce of eggs. It is the most useful and convenient method of transporting the insects about. (Fig. 13.)

Knife.—Constructed so as to cut the leaf easily and small. (Fig. 14.)

Double Chopper.—When the leaf is cut with the knife, it should be chopped fine with this instrument, to multiply the particles and edges of the leaf. This is only requisite in the first and second age. (Fig. 15.)

Large chopper—is made something like a straw cutter ; it is useful in chopping the leaf coarsely, and in great quantity. It is, used for the three first days after the third moulting. (Fig. 16.)

Small hand birch-broom.—This is used to spread the leaves evenly upon the hurdles. (Fig. 17.)

Small door, made in the large door.—At the bottom of the large doors, there should be a small door with a sliding panel to close at pleasure, and which acts as a larger species of ventilator. (Fig. 18.)

Square basket.—It should be wide and shallow, with a hook fixed to the handle, which may be hung on the edge of the hurdle or tray, and slide along the ledge backwards and forwards, without touching the edge of the lower hurdle beneath. (Fig. 19.)

Stepping-boards and benches—so constructed as to facilitate feeding the second row of hurdles conveniently, and without stooping. (Fig. 20.)

Short ladders—rather wider and more easy than

P

they are generally made, and sufficiently high to be made to lean against the hurdles. (Fig. 21.)

Barometer.—I have explained the use of it in a separate paragraph. (Fig. 22.)

Fumigating apparatus.—The glass vessel should be wide at the neck, more so than in those bottles which I described, (Chap. VII. § 2.) and instead of being closed with a cork, it should have a glass stopper. It should close hermetically, and it should be screwed fast with a vice or screw. This vessel, thus constructed, is more useful and convenient. (Fig. 23.)

Hotte, or *panier*—a basket for strapping on the back, to transport the worms of equal width from top to the bottom; of a close texture, to prevent any dung, &c., from dropping out. (Fig. 24.)

Box for dung and rubbish.—When the hurdles are cleaned, the rubbish remaining should be swept into this box; without this, it is not possible easily to clean the sheets of paper which are on the hurdles in the fifth age of silk-worms. (Fig. 25.)

Frames for placing the moths upon.—These frames should be covered with cloth, which may be changed when dirty. These frames may be used for carrying leaves in or about, and should have a handle such as that of the portable wooden trays. (Fig. 26.)

Box for keeping the moths in.—This should be a slight wooden or paper box, with air-holes in the sides; is very useful to keep the moths from light without injuring them, and to prevent the males from flapping their wings. (Fig. 27.)

Stand for the silk-worm eggs.—This is the most convenient invention I know for gathering the eggs; when it has been employed, it shuts up, takes up no room when put by till the following year, when it again comes into use. (Fig. 28.)

Twine frames.—There is no better manner of preserving the eggs, than hanging them on this; they receive the air on all sides, and the eggs are kept cool and dry. (Fig. 29.)

Chapter XIV.

GENERAL VIEW AND APPLICATION OF ALL FACTS STATED IN THIS WORK, AND WHICH ARE IMMEDIATELY CONNECTED WITH THE ART OF REARING SILK-WORMS.

When an art, so eminently allied to individual and national prosperity, is in question, an author cannot, I think, make the knowledge and practice of every branch of the art too clear, plain, and familiar.

If any persons who rears silk-worms have read this work but once, the observations I am going to make will spare their reading many of the chapters again ; the facts, when presented together, will besides make a deeper impression than when they are offered singly to the mind.

I do not doubt, that some of my readers will think many of the things I intend stating useless, because they offer no pecuniary result for individual interest, which is the first aim of the art of cultivating silk-worms.

But I must hope, that those who will bestow their attention upon the series of facts I present, will find an intimate, although perhaps not an obvious, connexion between them and those very results which the cultivator of silk-worms purposes to attain.

It may also appear strange, that I should have applied calculation to this art, with the same exactness and precision which might have been applied to certain and invariable objects of art ; whilst this art we treat of is subject to change, to the influence of numerous causes, and to alteration. But I thought it requisite to demonstrate the experiments reiterated for years, with the utmost exactitude, by the certainty and precision of calculation, that my practice might offer steady and sure rules to direct the cultivator of the silk-worms who might be inclined to adopt it.

This chapter will be divided into seven paragraphs.

1. Facts relative to the eggs of the silk-worms, and to their hatching.

2. Facts relative to the extent of space which should be occupied by the silk-worms, in their different ages.

3. Facts relative to the consumption of mulberry-leaves by the silk-worms, in their different ages, and observations on the subject.

4. Facts relative to the increase and decrease of the silk-worms in weight and size.

5. Facts relative to the cocoons containing the healthy chrysalis, the diseased chrysalis, or the dead chrysalis.

6. Facts relative to the production of the eggs.

7. Facts relative to the buildings and utensils.

1. *Facts relative to the Eggs of the Silk-worms, and to their Hatching.*

To compose an ounce of eggs of the largest breed of silk-worms of four casts, it will require 37,440 *.

* I must remark that the ounce here understood by the author, is smaller than the French ounce. From the information I could get, I found that 28 Lombardy ounces are about equivalent to 25 French ounces. It is from this, that I have reduced all the calculations into French measure. My reckonings may not always be exactly correct, but upon the whole, I believe they will be found free from any material errors.—*(French Translator.)*

If all these eggs produced a worm, and that all the worms lived, from one ounce of eggs, about 373lbs. of cocoons would be obtained, because 150 cocoons weigh about one pound and a half.

To form an ounce of eggs of common-sized worms of four casts or moultings, 39,168 will be required; if all these eggs produced a worm each, and all these worms lived, this ounce of eggs would yield 162lbs. of cocoons, because about 360 cocoons weigh a pound and a half.

To form an ounce of eggs of silk-worms of three castings or moultings, it will require 42,200 eggs.

If all these eggs produced a worm each, and all these worms lived, the ounce of eggs would yield 105lbs. of cocoons, because 600 cocoons weigh about a pound and a half.

From these positive facts, it may easily be ascertained, by the quantity of cocoons obtained, how many eggs have failed, and how many have died in the various ages; it will afterwards be of use in determining which method of rearing the worms is most favourable to their preservation.

From the time of laying the eggs, until they are taken off the cloths, that is, a period of nine months, they only lose about $\frac{1}{100}$ of this weight.

From the day the eggs of the common silk-worms of four casts are put into the stove-room, until the time they begin to hatch, they lose, on an average, 47 grains per ounce, which is equivalent to $\frac{1}{12}$ of their total weight.

The weight of the shells of the eggs, after the hatching of the worms, amounts to 116 grains per ounce, which is about $\frac{1}{5}$ of the total weight.

Consequently, the loss of weight of the eggs in the stove-room deducted, and the weight of the shells, 54,625 silk-worms just hatched make an ounce, while to make that weight 37,168 eggs were sufficient.

Thirty-nine thousand silk-worms, proceeding from one ounce of eggs, can eat the first day, and lie easily in a space of about 20 square inches.

2. *Facts relative to the Extent of Space which should be occupied by the Silk-worms in their different Ages.*

The worms proceeding from one ounce of eggs should have a space,—

In the first age, of seven feet four inches square.

In the second age, of fourteen feet eight inches.

In the third age, of thirty-four feet, six inches square

In the fourth age, of eighty-two feet six inches square.

In the fifth age, of one hundred and eighty-three feet four inches square.

As the silk-worm rises in the fifth age, I would here willingly state the weight of the materials

required to form the hedges and cabins sufficient
for receiving the worms of one ounce of eggs;
that is to say, 120lbs. of cocoons; but I have not
been able to fix any certain rules, as the varieties
of brush-wood, haulm, straw, heath, or vegetable
substances used for the purpose, differ so much in
weight, that the calculation would scarcely be
correct ; 150lbs. of haulm weigh as much as
450lbs. of heath, and that again as much as
750lbs. of broom.

3. *Facts relative to the Quantity of Mulberry
Leaves consumed by the Silk-worms in their
different Ages. Observations on the Subject.*

The result of the most exact calculations is, that
the quantity of leaves drawn from the tree, employed for each ounce of eggs, amounts to one
thousand six hundred and nine pounds eight ounces,
divided in the following manner :—

	lbs.	
First age, sorted leaves	6	
Second age, ditto ditto	18	lbs.
Third age, ditto ditto	60	1362
Fourth age, ditto ditto	180	
Fifth age, ditto ditto	1098	

Per ounce of Eggs, sorted leaves   1362 *

* I have given some idea in the course of this work of what
I mean by *sorted leaves*, but I will further explain. Great
care must be taken in picking and sorting the leaves for the
feeding of the worms of the first ages, such as picking off all
the twigs, stalks of the leaves, spots,&c., and to clear them as

<table>
<tr><td></td><td>lbs.</td><td>o</td></tr>
<tr><td>Brought over</td><td>1362</td><td>0</td></tr>
</table>

But this leaf has lost by sorting so much weight, in the following proportion :—

Refuse picked from the leaves.

	lbs.	oz.		
First age	1	8		
Second age	3	0		
Third age	9	0	142	8
Fourth age . : . . .	27	0		
Fifth age	102	0		

Refuse picked off, per ounce of eggs 142 8

Total 1504 8

During the whole period of rearing the silk-worms, the 1609lbs. 8oz. of the leaves drawn from the tree, have lost by evaporation, and other causes, besides sorting and picking, as above stated   105 0

Total 1609 8

much as possible from all useless parts. This operation is most essential in the two first ages, when the leaves are to be chopped very small.

In the third age, the sorting and picking the leaves is not of much consequence, and still less so in the fourth age, and fifth age.

The sorting and picking is of importance, inasmuch as it enables you to put 15 or 20 per cent. *less* substance upon the wickers than would otherwise be done, and which the worms do not eat. This substance increases the litter and the moisture, without necessity or motive ; in the climates, where the worms are in the open air, it would of course be unnecessary to sort the leaves.

In the fifth age, and even in the fourth, when the season is favourable, leaves mixed with a quantity of mulberries, boughs and stalks, may be put on the hurdles, although it is known that the worms do not eat them, because at that period it

It has been observed that the leaves distributed on the hurdles per ounce of eggs, was 1362lbs.

During the life of the silk-worms, there has been carried away from the hurdles in dung,

	lbs.	oz.
In the first age	1	4
In the second age	4	8
In the third age	19	8
In the fourth age	60	0
In the fifth age	660	0
Total	745	8

The excremental substance found in the litter, or in the fragments of uneaten leaves, when taken off the hurdles, weighed

	lbs.	oz.	drs.
First age	0	1	4
Second age	1	3	0
Third age	3	9	4
Fourth age	18	9	4
Fifth age	132	0	0
Total	155	7	4

would be too troublesome to sort so large a quantity perfectly, nor is there the same motive to do so. These substances being by this time grown large, hard, and woody, are less liable to fermentation, although they may accumulate as litter. If the laboratories are constantly dry, and well aired, these substances will do no mischief, but keep the litter light, and allow the air to circulate more freely through it.

When the silk-worms find any leaves they do not like they leave them. There are some of a hazel dark colour, which have fermented slightly; these the worms will eat, if they are not quite spoiled, nor could I ever perceive they were the worse for it; from which fact it would appear the fermentation had not affected the saccharine or resinous part of the leaf.

Deducting 155lbs. 7oz. 4drs. from 745lbs. 8oz., there will remain 590lbs. 4drs. of vegetable substance, namely, in stalks, fruit, fragments of leaves, &c., not eaten by the silk-worms; and subtracting the 590lbs. 4drs. of leaves from 1362lbs. that had been laid on the hurdles, it will appear, that the worms have only really consumed 771lbs. 7oz. 4drs. of pure leaves.

From the above statement it ensues,

1. That to obtain a pound and a half of cocoons, it requires about 20lbs. 4oz. of leaves, as gathered from the tree: and that it requires 1609lbs. 8oz. to obtain 120 pounds of cocoons, which an ounce of eggs should yield.

2. That this quantity of leaves gathered from the tree, deducting 142lbs. 8oz. of refuse and sorting, and 105lbs. of decrease, by means of evaporation, it only requires 16lbs. 8oz. of pure leaf, per pound of cocoons, or 1362lbs. for 120lbs. of cocoons.

3. That subtracting from the 1362lbs. the 590lbs. 4drs. of residue, such as little branches, stalks, fruit, &c., which were taken off the hurdles with the litter, 9¾lbs. of pure leaf have been sufficient to obtain 1½lb. of cocoons, and consequently 771 lbs. of leaves effectually eaten have sufficed to obtain 120lbs. of cocoons.

4. That the 1362lbs. of leaves distributed on

the hurdles, having only yielded 745lbs. 12oz. of dung, including excremental substance, and 120lbs. of cocoons, in all making 865lbs. 12oz., there is a loss escaped in gas, vapour, and steam, &c., in the laboratory, of 496lbs. 4oz.

5. That three parts nearly of these 496lbs. 4oz. of substance having been extracted, as we have shewn (in Chap. VIII.), in the last six days of the fifth age, it follows, that in those days, the above substances weighed daily 30, 40, and 50lbs.

6. That in a laboratory containing the worms proceeding from five ounces of eggs, such as we have described, there must have escaped on each of the last six days of the fifth age, 300 or 450lbs. of gas and vapour, invisibly to the eye.

These latter statements which I mentioned else-where, and which would seem incredible were they not demonstrated by exact calculation, thus connected, present strong evidence of how formi-dable the enemies are which assail the laboratory.

These evils are unknown in the hot climates whence originated the silk-worms, because there these insects are always in contact with the external air, which circulates freely, and dispels gases and mephitic vapours.

Although the cultivators among us know not the force of the material cause that produces the death of their silk-worms, they, however, know that in the last period or age, every part of the laboratory should be opened ; but often in avoiding one danger, they meet another, such as exposing the worms to cold and wind, which may harden them, and cause them to drop off at the moment they had begun to weave the cocoon.

There is only the gentle and continual renewal and motion of the internal air that can be beneficial and natural to the silk-worm.

It is considered astonishing, that one single worm which, when first hatched, only weighs the hundredth part of a grain, should consume, in about thirty days, above an ounce of leaves, that is to say, that it devours, in vegetable substance, about 60,000 times its primitive weight.

The result of my experiments tends to shew, that in warmer climates than ours, the silk-worms consume rather less leaf than I have here stated, because the quality of the leaf is more nutritive.

In the favourable regions of Dalmatia, I obtained in 1807 one pound and a half of cocoons from fifteen pounds of leaves, and fifteen pounds of cocoons yielded one pound and a half of silk, although it was not so delicate and fine as ours. Notwithstanding the richness of the produce in that province, on which nature has lavished so much, there are few mulberry plantations to be found in it.

4. *Facts relative to the Increase and Decrease of Silk-worms, in Weight and Size.*

PROGRESSIVE INCREASE.

	gr.
Hundred worms just hatched, weigh about .	1
After the first moulting	15
After the second moulting	94
After the third moulting	400
After the fourth moulting	1628
On attaining the greatest size and weight . .	9500

Thus have they in thirty days increased 9500 times their primitive weight.

	lines.
The length of the silk-worms when just hatched is about	1
After the first moulting its length is	4
After the second moulting	6
After the third moulting	12
After the fourth moulting	20
After the fifth moulting many attain the length of .	40

The length of the silk-worm is thus increased forty times in twenty-eight days.

PROGRESSIVE DECREASE.

	grs.
100 silk-worms when arrived at the highest state of maturity, size, and perfection, weigh . .	7760
100 chrysalides weigh	3900
100 female moths weigh	2990
100 male moths	1700
100 female moths having deposited their eggs .	980
100 female moths dying naturally, after having laid the eggs, and nearly quite dried . . .	350

In the space of twenty-eight days more, the silk-worm has diminished, or lost, thirty times its own weight.

Its length, from the period of the largest growth until it changes into the chrysalis, diminishes about two-fifths.

During and immediately after coupling, the moth appears to augment in weight; 100 moths, which before coupling weighed 2990 grains, weigh immediately after 3200 grains. This is caused by the weight of the matter injected by the male, which is heavier than all the female had lost at the same time.

The worm diminishes gradually in weight during the last twenty-eight days of its existence; that is, from the moment of attaining its perfection as a worm, until its death in the form of a moth, it eats nothing, is supported by its own substance, and yet accomplishes in that period the most important functions of its life.

The facts I have stated demonstrate the strong vitality of the silk-worm, and what pains and errors must be adopted to disease and kill it.

5. *Facts relative to the Cocoons containing the healthy Chrysalis, the diseased Chrysalis, and the dead Chrysalis.*

When the cocoons are perfectly formed, they diminish, in the four first days, three-quarters per cent. each day; the other days the diminution is very trifling.

^{ozs.}

	ozs.
1000 ounces of perfect cocoons composed of living chrysalides	842
Of the remains or envelopes cast by the worms when they become chrysalides	4½
Of pure cocoon	150½
Total . .	1000

Each healthy cocoon proceeding from a well-managed laboratory contains the seventh part, and even $\frac{2}{13}$ of pure cocoon, when compared to the weight of the cocoon containing the chrysalis.

However, the fact is, the average quantity of the cocoon obtained by the silk-mill is about $\frac{1}{12}$ of spun-silk, or 140 ounces of perfect cocoons, while the healthy chrysalides, which contain about twenty-one ounces of pure cocoon, will generally only produce twelve ounces of silk.

Let us now connect the facts I have stated. About 97 pounds 8 ounces of mulberry leaves, will produce 7½ pounds of cocoons; that 7½ pounds of cocoons containing the healthy chrysalides yield about 18 ounces of pure cocoon; that these 18 ounces of pure cocoon only give eight ounces of spun silk.

The proportion, then, between the weight of the mulberry leaves and that of the pure cocoon, is something about 87 to 1; and the proportion of the weight of the mulberry leaf and that of the spun silk, is of 152 to 1.

When the proprietor, therefore, has grown 228 pounds of leaves from the mulberry tree, he has contributed to the production of $1\frac{1}{2}$ pounds of spun silk, and some ounces of the coarse floss, as we shall show presently.

The proportion between the spun silk drawn from the cocoon and the cocoon itself, may vary according to the ill or good management of the worms.

In the year 1814, which was unfavourable, my cocoons yielded me about 15 ounces of very fine silk from $7\frac{1}{2}$ pounds of cocoons; I even obtained 13 ounces from $7\frac{1}{2}$ pounds of the inferior refuse cocoons.

The proportion between the weight of the cocoons containing the healthy chrysalis which can be spun, and that part called the coarse floss which cannot be spun in the same manner, is, on an average 19 to 1; that is to say, we find one pound of coarse floss to nineteen of cocoon that can be spun.

What I have stated shews that the weight of silk floss and chrysalis, proceeding from a given quantity of cocoons, is not equivalent to the weight of the cocoon itself. The reason of this is, that there are two other substances contained in the cocoon; one which the silk-spinners work up and sell at a low price, and another which, being of a gummy nature, dissolves and is lost in the cauldron.

The proportion between the quality of spun silk obtained from the cocoon and that of the floss or above-mentioned, is on an average of about from 110 to 40, or 11 to 4; that is to say, that there are four ounces of floss to eleven ounces of silk.

Generally, in 150 pounds of cocoons there will be found about $1\frac{1}{2}$ pounds of double cocoons, formed by two worms, and are worth not quite half a single cocoon.

About 506 feet of the single thread of silk, spun and extracted from silk-worms of three casts, weighs one grain!

The cocoon of the small worm of three casts yields $2\frac{304}{}$ grains of silk; and if we make an average calculation, we shall extract about 11 ounces of silk from 3000 cocoons, weighing $7\frac{1}{2}$ pounds.

The same cocoon yields about 1166 feet long of the single thread; in one ounce of this spun silk will be found a length of 291,456 feet; 458 feet 4 inches of spun silk, extracted from a common cocoon of four casts, weighs one grain.

This common cocoon of the silk-worm of four casts yields $3\frac{84}{100}$ grains of silk; because, on an average, there are about 11 ounces of spun silk

drawn from 1800 cocoons, which weigh $7\frac{1}{2}$ pounds.
This cocoon yields 1760 feet of spun silk; the
ounce of this spun silk is 264,000 feet long:
421 feet 8 inches of the silk spun from the great
cocoon of large worms of four casts, weigh a grain.
This same cocoon yields $9\frac{216}{100}$ grains of spun silk;
because, on an average, 11 ounces of silk are
drawn from 750 cocoons, which weigh $7\frac{1}{2}$ pounds.
This cocoon consequently gives about 3885 feet
of spun silk; one ounce of this spun silk is
242,880 feet long.

It appears astonishing that the silk-worm in
about three days, which it employs in forming
the cocoon, should produce such an enormous
length of silk; and in this I do not include the
first down which is taken off the cocoon, nor the
coarse floss.

We may conclude, on an average, that the silk-
worm in forming the cocoon draws a thread of
half a mile in length*.

The proportions between good cocoons that
have been pierced, and have been used for the production of eggs, and the remains which they contain, vary little. These cocoons cannot be spun,
because the continuity of the thread has been
broken by the moth. The empty cocoons are

* In l'Abbé Rozier's *Cours d'Agriculture*, it is stated
that one single thread that has formed a whole cocoon is three
miles long.

always soiled inside and outside ; and even when perfectly dry, they retain a sort of stain.   They never are so clear as those cocoons which have been cut when the chrysalis was alive, and consequently they are heavier.

That nothing may be concealed from those that wish to rear silk-worms, I will here state the proportionate difference which may be observed in the pierced cocoons from which moths have issued :—

	ozs.
1000 ounces of these cocoons only weigh about	170
The remains or envelopes of the worm which is become chrysalis	$5\frac{3}{4}$
The remains of the chrysalis which the moth leaves on issuing from the cocoon	$7\frac{1}{4}$
	183

The 1000 ounces of cocoons chosen for eggs have, then, yielded rather more than the sixth part of the weight of the empty cocoon ; they weigh 170 ounces, when the 1000 ounces of cocoons with the empty chrysalis only yielded 153 ounces of pure cocoon.

Before I conclude my observations on the healthy chrysalis, I shall quote a fact that may appear surprising.—It requires 12,860 cocoons to form 1000 ounces.   It has been seen that the remains or envelopes of this same quantity of silk-worms, weigh about $4\frac{1}{2}$ ounces.   Let us suppose that the worm, at its highest degree of size and growth, should, on an average, be only three inches long, and nine lines in circumference, the skin must be $2\frac{1}{4}$ inches, square surface ; the 12,860 skins must therefore offer a surface of 28,935 square inches. This surface is equal to 110 feet, and does not weigh above $4\frac{1}{2}$ ounces. Knowing so exactly the various proportions of the cocoons containing the healthy chrysalis, we shall see how different are those of the calcined or decayed chrysalis.

These proportions should be known, because they are immediately connected with the art of rearing silk-worms with the utmost advantage to the proprietors.

The ideas entertained on the subject of the cocoon containing a calcined worm, are generally confused.

Many cultivators complain of the loss they sustain in selling the light cocoons to the silk-spinners at a low price ; and the purchasers deny deriving any considerable advantage from this purchase.

The purchaser may not always be wrong, but still the losses of those who sell are heavy and real ; and it is difficult to understand how they could for years have been satisfied with complaining, instead of seeking to discover and overcome the cause of these losses. (Chap. XII.)

Cocoons of the Calcined Worms without Stain*.

ozs.

1000 ounces of these cocoons contain in chrysalis,
dried worm or mummy, and envelopes of dry
saline substance   642
Pure cocoon   358

Total . . 1000

The proportion between the weight of the mummy and the pure empty cocoon is 18 to 10.

Seven pounds and a half, that is to say, 120 ounces of these cocoons, contain about 44 ounces of pure cocoon.

As these cocoons are not stained, and yield about 12 ounces of spun-silk, out of 21 ounces of pure cocoon, it is evident that from 500 ounces of pure cocoon may be drawn,—

ozs.

In spun silk about   .   .   . 28½
In coarse floss or other substances . 21½

Total . . 50

If the spinner only obtains twelve ounces of silk from 7½ of the healthy chrysalides, while he can get twenty-five ounces from the same weight of calcined unspotted cocoon; in buying these, there are thirteen ounces more of silk in the 7½ of

* I should think the author meant the white comfit chrysalis *(dragées),* or calcined cocoon.—*Translator.*

cocoons, that is to say, double as much. Consequently, if the common cocoon costs, for instance, three francs a pound, those that are calcined and unstained should cost six francs a pound.

There should be about 1100 of the calcined worms to form a pound and a half*.

Stained Cocoons containing the Calcined Chrysalis.

One thousand ounces of these cocoons contain
a weight of mummy chrysali, with saline
substance, amounting to   . . . .   600 ounces.
Pure cocoon   400

                       Total   1,000

The proportion of the full cocoon, and empty, must then be three to two.

* I found the weight of the empty cocoon of the healthy chrysalis, nearly of the same weight as that from which I had extracted the calcined mummy.

The spinner, however, often finds an impediment which prevents his drawing from the calcined quality of the cocoon, the same quantity of silk ; this impediment is caused by the lightness of the cocoon, that has the calcined worm. It is almost indispensable, when spinning the cocoon, that there should hang a weight, such as the healthy chrysalis to the end of the thread, to keep it steadily down in the water. If the chrysalis is calcined, it weighs less than the healthy one, and the spinner will find it a great disadvantage ; it will not keep in the water, nor spin steadily, and the spinner gets rid of it as soon as possible.

This will show the benefit of healthy chrysalides in all the cocoons, as they weigh six or seven times more than the empty cocoon.

Seven pounds and a half of cocoons, that is to say, 120 ounces, contain about fifty ounces of pure cocoon, but as in all stained cocoons there is always a part of the substance affected and spoilt, the spinner cannot foretel whether from seven pounds and a half of cocoons he will obtain one half of the quantity that the healthy cocoons would yield him; the less the silk the greater will be the quantity of coarse floss, and the floss is worth less than the cocoons of the healthy chrysalis.

A thousand of these stained cocoons weigh a pound and a half.

Cocoons of which the Chrysalis is Decayed, or Gangrened, Stained, and Unstained.

Generally, it is not possible to separate these chrysalides from the cocoon; the worm or chrysalides being turned into a corrupt black soapy substance sticking to the inside of the cocoon. Sometimes the mummy is exceedingly black, and now and then detached, and most frequently it adheres to the cocoon.

A part of these cocoons may be spun; the stain does not always alter and spoil the silk. The spinners can never be sure of the quantity of silk they may be able to wind from them, and they in general dislike working on these cocoons, although many speculate and imagine it profitable, buying them at a low rate.

The silk drawn from diseased cocoons is never so fine as that given by the perfect and healthy cocoon.

Eight hundred and sixty cocoons of the black chrysalisweigh a pound and a half.

The cultivator thus loses two-thirds, or three-fifths, upon this quality of cocoon.

I must here state, in speaking of these three sorts of cocoons, that I have only made my experiments upon such cocoons as were brought to me from various places; therefore my calculations may very possibly differ from those of other observers.

I will close this paragraph by the following remark. The art of spinning or winding the silk from the cocoon, is as yet entirely in the hands of people as ignorant as those who have hitherto reared the silk-worms.

For instance, it is a well-known fact, that of two spinners, spinning each 7½lbs. of cocoons of the same quality, one will extract constantly eight ounces of silk, whilst the other will only draw six ounces and a half, and perhaps less still. There are even spinners so ignorant, that in giving constant jerks and strokes of the handle, they destroy several layers of silk enveloping the cocoon:

Q

others extract less silk, because the water is too hot in which they spin.

Much is annually lost, by the awkwardness and carelessness of the spinners.

6. *Facts relative to the Production of the Eggs.*

Three-hundred and sixty cocoons, of the finest quality, weigh about 25 ounces.   If we suppose half of these to be females, these will be about 180.

Each impregnated moth weighs about 32 grains, and altogether 5740 grains, which make about 10 ounces.

After four, five, or six days, each moth will have laid, on an average, 510 eggs.

This number of eggs is equivalent to $7\frac{1}{2}$ grains, as 68 eggs weigh a grain.

The 180 female moths consequently lay 91,800 eggs, which weigh 1350 grains, or about two ounces and one-third.

This proportion of two ounces one-third per pound of cocoons, augments and diminishes, according as in the 360 cocoons which form one pound and a half, females predominate, and *vice versâ*.

At the end of four days, the 180 moths that have laid their eggs only weigh 1800 grains.   As it has been stated, that the eggs weighed 1350 grains, it will appear that the moths have lost, in four days, 1,610 grains in earthy, liquid, and aeriform substance.

If the 91,800 eggs, obtained from 180 moths, yielded an equal number of silk-worms, and that, well managed, they each in time formed a cocoon, from the eggs produced by the above-mentioned 1½lb. of cocoons, we should obtain 382lbs. eight ounces of cocoons, which the following year would yield eggs sufficient to produce 97,537lbs. eight ounces.

7. *Facts relative to the Buildings and Utensils.*

To make the laboratory of the silk-worms similar to their native climate, they must be enabled to live in it perfectly free from all moisture ; the temperature should neither be too hot nor too cold, and above all, they should never be exposed to sudden transitions from one state of atmosphere to another ; the air should always circulate gently.

A warehouse, a cellar, or any other low, shady, close place, is best calculated for the preservation of the leaves during two or three days, provided it be cool, damp, and shut from air and light.

The utensils employed to rear silk-worms, are constructed so as to spare time and expense, and for the better management of the silk-worms and the eggs at all times. The interest of the cultivator, and the progress of the art, make the rules which I have laid down indispensable.

Chapter XV.

OF THE NATIONAL AND INDIVIDUAL ADVANTAGES WHICH WOULD ACCRUE TO THE PUBLIC, THE PROPRIETORS, AND CULTIVATORS, BY IMPROVING THE METHOD AT PRESENT ADOPTED FOR THE REARING OF SILK-WORMS.

THE European soil offers a given number of natural productions, everywhere the same, and everywhere indispensable.

The favourable or unfavourable changes which annually affect these productions, cause whole nations alternately to buy and sell, as often occurs in corn and other grain, &c. &c.

Political calculation and financial interest regulate in different states the exports, imports, and customs, according to circumstances.

But when nature, favouring the soil, has endowed it with a faculty of yielding constantly, and in an indefinite degree, a production far exceeding its own wants, and necessary to those of other countries, or required by their luxury, political interest, and financial calculation should become stable and liberal, as nature is in bestowing the production, and the maxim which ought to guide administration should be couched in simple terms.

" Encourage the cultivation of the production; " protect exportation; and act so as to secure its " free circulation in all foreign markets, that the " consumption of it may increase abroad."

In this manner would the policy of states become connected with the interests of those states.

This maxim should be fundamental in Italy, in everything relating to the silk-trade, as by its annual value, it should be placed immediately after our more important production, corn and wine; and it is even of superior value to those as a more *exportable* commodity for foreign markets.

The value of exportable silk is double the amount of any of our other productions. Besides, there is not in the European markets any production which, compared with its own intrinsic value, offers a larger net profit than that which silk yields. By natural value, I understand that which results from a combination of values, such as revenue of the capital produced by the mulberry-tree, and the interest of the advances made for obtaining silk, and the amount of all salaries paid.

Notwithstanding this, it is demonstrated, that silk is yet far from having attained its utmost degree of value, principally owing to its defective cultivation, and the imperfection of the art of rearing silk-worms, and the errors constantly committed by the various Italian administrations.

During late years, the government which has lately ceased, by a confusion pervading all ideas of political economy, thought fit to load the silk-trade with enormous taxes, customs, monopolies, prohibitory systems, &c. It appeared as if the intention had been to diminish production by impeding exportation; reason, calculations of experience, and the knowledge of enlightened and scientific statesmen, were all in vain, and had not power to destroy this absurd system.

I had occasion, in 1812, to explain, in the general council of arts and commerce, my ideas relative to the importance of this commodity.

These ideas, which were adopted by my colleagues, produced no beneficial advantage, although represented to the government. Thus it was evident, that by erroneous calculations, the views of administration were in direct and manifset opposition to the dearest interests of the state.

There is no enlightened nation which does not employ every means to carry its surplus of annual produce to foreign markts, and there cannot exist an intelligent administration which would not facilitate this beneficial operation. In England, for instance, the administration of finance will return the duties on foreign produce, when this produce is to be exported after being worked or manufactured. It is upon this simple principle of political economy, and other circumstances, favourable to the industry of that nation, that it is enabled to strike at the industry of other countries, which, not enjoying the same advantages, cannot compete with it in foreign markets.

1. *Annual Value of the Cocoons, or of the Silk obtained from them, and exported.*
Some Observations on the Value of the Manufactured Silk which may be exported.

I here offer a sketch of the value of silks and other produce of the cocoon, which have been exported in the latter years of the kingdom of Italy, which has just ceased to exist. I shall add a note of the other articles more or less manufactured from the cocoon, and the silk exported.

The exact quantity exported may be known by the Custom-House registers, where may be found the amount of duty paid.

As to the value of the articles, it is determined, by the declaration of merchants and manufacturers, from current prices.

There may therefore be differences, at least in the specified valuation, but in nothing else; and as the system of the customs was vexatious, aggravating, and contrary to the best national and commercial interests, smuggling was frequent, and, perhaps, from circumstances, unavoidable. It was carried on under the eyes of every one. For which reason I shall add to the quantity of silk exported $\frac{1}{15}$ per cent., to be nearer the real value of exportation.

It is painful to be obliged to comprehend in this calculation the contraband exportation consequent on the errors of administration; but it will ever be so, whilst the Custom-House regulations are in opposition to national interest. When these regulations accord with national interest, contraband ceases; each branch of national industry is guided and animated by individual interest; production and consumption increase; the crimes and immoralities attendant on smuggling disappear, and all speedily return to order and tranquillity.

It cannot be understood why, in almost all places, under various pretexts, the exportation of *raw* silk has been impeded and fettered.

Some administrations, to increase the salaries and profits, for instance, of the country which grew the silk, by two francs per pound of spun silk, have put an excessive duty on the exportation of that species of silk; some have gone so far as to prohibit exportation altogether. From this it

often happened that, to gain two francs, the speedy
means of selling the silk at 28 or 30 francs a pound
were foregone, and thus consumption and competition were considerably diminished.

It may be easily conceived that many foreign
purchasers often prefer working the raw silk after
their own fashion, when we have seen them buying raw silk in our markets dearer than silk that
had been spun.

It may be useful to lay high duties on exportation of primitive articles used for manufactures,
when, having been worked in the country which
produced them, they may be preferred in foreign
markets; but these duties are most pernicious to
the national interest, when they diminish exterior
consumption and lessen the demand. Unfortunately it is always easier in political economy to
adopt errors than to correct them. And we have
often seen these usurping the place of the liberal
and beneficial truths of rural, mechanical, and
commercial industry.

We must hope that time will overcome them.

In the present state of things, the value of exportation of raw and manufactured silk would
surprise the reader.

If the kingdom of Italy exported, in average
years, silk to the amount of 83 millions of francs,
and if, during some years, as I shall shew, the value
of exportation was beyond 110 millions, it is in-

dubitable that it is capable of vast increase, by merely improving the cultivation of the silk-worms, as I have indicated, and in the plantation of the mulberry-tree, as I intend shewing shortly.

APPENDIX.

THE sketches I here offer may give some idea of how great the value of that production is which nature has bestowed on Italy.

Exportation of Silk, and Articles connected therewith from Italy.

1807.

	lbs.	Milan livres†.
Raw silk*, Milan pound of		
12 ounces . . . 137,518	2,475,324	
Spun silk 3,038,372	42,805,812	
	45,281,136 ⎫	52,073,306
Augmentation of 15 per 100 . .	7,792,170 ⎭	
Dyed silk . . . 255,367	7,607,754	
Coarse silk (or *Filoselle*) . 80,100	220,275	
Floss 74,100	111,160	
Coarse Floss . . . 721,100	273,384	
Silk stuffs 179,331	12,620,490	
Mixed ditto . . . 1,069	47,180	
Floss ditto 9,961	249,025	
Gauze and veils . . . 30,311	2,727,990	
Sewing silk . . . 5,332	243,094	
Ribands, silk, and mixed . 23,586	909,540	
Ditto floss . . . 7,858	196,450	
Sundries 1,051,612	26,257,944	
Total . .	78,331,250	

* As I before observed, the 28 ounces of Milan are equivalent to about 25 French ounces.

† Six livres, ten sous, three deniers of Milan, are worth five French francs.

	lbs.		Milan livres.
		Brought up,	78,331,250

1808.

		lbs.	
Raw silk	233,378	2,800,536	
Spun silk . . .	.3,127,492	31,912,380	

	34,712,916 ⎰	39,919,853
Augmentation of 15 per 100 . .	5,206,937 ⎱	

	lbs.		
Dyed silk	244.282	5,200,211	
Coarse silk	93,400	186,800	
Floss . , . .	101,400	116,610	
Coarse floss . . .	801,860	235,882	
Silk stuffs	220,551	1,195,448	
Mixed ditto . . .	2,949	103,120	
Floss ditto . . .	11,588	222,489	
Gauze and veils . . .	29,761	2,053,509	
Ribands, silk, and mixed .	21,279	643,580	
Ditto floss	8,349	160,300	
Sewing silk . . .	4,896	150,258	
Sundries		323,699	10,591,906

1809.

		lbs.	
Raw silk . . .	310,358	3,724,296	
Spun silk . . .	.2,310,576	34,658,640	

	38,382,936 ⎰	44,140,376 .
Augmentation of 15 per 100 . .	5,757,440 ⎱	

	lbs.		
Dyed silk	225,800	4,840,162	
Coarse silk	62,700	125,400	
Floss	163,000	187,450	
Coarse floss . . .	765,700	226,374	
Silk stuffs	179,487	9,725,004	
Mixed ditto . . .	2,061	72,923	
Floss ditto	8,142	156,326	
Gauze and veils . . .	18,609	1,284,021	
Ribands, silk, and mixed .	7,392	216,857	
Ditto of floss . .	9,581	183,955	
Sewing silk · . .	4,013	132,421	
Sundries		306,351	17.457,248

		190,440,633

1810.		lbs.		Milan livres.
			Brought up,	190,440,633
Raw silk, new weight	.	153,286	5,763,553	
Spun silk . . .	.	826,784	46,630,617	52,394,170
			52,394,170	
Augmentation of 15 per 100	. .		7,859,125	60,253,295
Dyed silk, new weight	.	113,015	7,943,373	
Coarse silk . .	.	37,000	242,734	
Floss · . .	.	63,800	239,437	
Coarse floss . .	.	309,600	188,009	
Silk stuffs . .	.	70,692	11.739,135	
Mixed ditto . .	.	306	33,158	
Floss ditto . .	.	3,482	204,765	
Gauze and veils .	.	13,302	2,809,466	
Ribands, silk, and mixed	.	4,705	405,934	
Floss ditto . .	.	2,290	134,681	
Sewing silk . .	.	2,149	211,761	
Sundries, other silken articles	.		390,690	24,543,143

In four years, 325,631,241

This table will be sufficient to shew, without any further observation of mine, the immense value of these annual exportations.

In 1810, the value of raw, spun, and dyed silk alone, amounted to eighty-five millions. This fact will demonstrate the produce of silk-worms to be such a source of riches, that were it to fail one year only, the failure would prove a national calamity.

2. *Annual Profit which the Proprietors and Cultivators may obtain in the Cultivation of Silk-worms, when the Proprietors furnishing the Leaves, and the Cultivators giving their Labour, they divide the Cocoons they obtain.*

In a work which I published in 1806, I spoke of the

necessity of creating new branches of industry among us : I pointed out the importance of the produce of cocoons, and the necessity of augmenting it.

I thought, when a general peace was proclaimed, and maritime commerce was free, expecially that connected with the Black Sea, that we should with difficulty encounter competition in the foreign corn markets. I stated my motives. This object appeared to me worthy the inquiry of the politician and the philosopher.

I had not then thought much upon the production of cocoons, because I had not personally entered into the cultivation of the silk-worm. However, my assertions were founded, at that time, upon an important series of facts.

I can now demonstrate in what the gain and profit of the proprietor and cultivator consist ; if they both attend to the proper culture of the mulberry-tree, and cultivation of the silk-worm.

It is a certain fact, that if these insects are well managed, 21 pounds of mulberry-leaves will be sufficient to obtain a pound and a half of cocoons (Chap. XIV.); 21,000 pounds of leaves will then yield 1500 pounds of cocoons, 750 of which should belong to the proprietor, and 750 pounds to the cultivator.

The 750 pounds of the proprietor will cost him the rent of the ground occupied by the trees that yield the leaf, and the interest of the capital employed in the advances made for having these mulberry-trees.

The 750 pounds of the cultivator cost him the price

of his labour and time, and some trifling expenses, which we shall mention hereafter.

As to the proprietor, we suppose him, in one instance, to have had, for a length of time, a sufficiency of mulberry-trees on his own lands, to obtain the given quantity of leaves ; and, on the other hand, we suppose the proprietor does not possess trees enough, and is desirous of increasing them.

To obtain 21,000 pounds of mulberry-leaves, the proprietor need only have sixty grafted mulberry-trees, each producing $7\frac{1}{2}$ pounds ; sixty trees producing 15 pounds ; sixty producing $22\frac{1}{2}$ pounds ; sixty trees producing 30 pounds ; sixty, producing $37\frac{1}{2}$; as many producing 45 pounds; sixty producing $52\frac{1}{2}$ pounds; sixty producing 60 pounds; sixty producing $67\frac{1}{2}$ pounds ; and ten trees producing 75 pounds ;—in all, 550 mulberry-trees. And as we must take into our consideration, that in these climates only one fourth part of these trees can be stripped every year, and consequently rests one year, it will require 732 trees.

A property which has for some years afforded 732 mulberry-trees of various sizes, such as I have described, and in which more are planted every now and then, to replace those that die, will yield 21,000 pounds of leaves.

There should be round the foot of a thriving mulberry-tree a fallow space of ground, of about four feet square, for many years, that the roots may receive external air, and draw the nutritious moisture

which the soil affords. Seven hundred and thirty-two
mulberry-trees will cover about 2928 square feet of
land, which cannot be sown or made any other use
of. If the land be good, on which the trees stand,
such as would yield a quarter and a half of wheat, or
twelve bushels, it would be, on an average year, a
loss of sixteen francs.

Next, suppose that the price of the mulberry-trees,
and other costs of plantation, come to two francs ;
732 mulberry-trees will cost 1464 francs, the interest of which sum is about 73 francs.

As we must allow a loss of four per cent. on the
mulberry-trees that die, we should add about fifty-
nine francs, which must be reimbursed to the proprietor before he clears his account.

From this calculation, the proprietor ought to obtain
annually,

	Francs.
Ground rent	16
Interest of capital employed . .	73
Casualties, loss and re-planting of the trees	59
	148

For the value of this sum, the proprietor will receive each year 750 pounds of cocoons. Supposing
he had no mulberry-trees on his land, and was
obliged to plant, to obtain the required quantity of
leaves, he would plant 1000 trees, wishing to keep
up that number. The ground employed could bear
nothing else, and would be of about 3600 square feet,
equal to a loss of eighteen bushels of wheat, 120

pounds weight, worth 24 francs. One thousand mulberry-trees planted would be worth 2000 francs capital, the interest of which would be 100 francs.

The annual loss on young mulberry plants is calculated at three per 100, consequently the value disbursed for 30 plants is 60 francs. Therefore the proprietor should obtain

	Francs.
Ground rent	24
Interest of 2,000 francs for planted trees	100
Casualties and replacing trees . . .	60
Total . .	184

The good and most advantageous mode of culture requires that the mulberry-trees, when transplanted, should not be stripped for three years, and that the fourth year they should be thinned in the branches and pruned; the fifth year they will produce an abundance of leaves, and may be stripped without danger.

It is, however, more beneficial not to strip them till the sixth year. After this first stripping of the leaves, they are cultivated according to the best common method, which is well known.

From what I have said, it must clearly appear that the proprietor loses 184 francs annually for four years without return.

At the end of the sixth spring, 1000 mulberry trees will, on an average, yield each 12 pounds of leaves; therefore there would be produced 12,000 pounds of leaves proper for the food of silk-worms, these 12,000 pounds of leaves should produce 855 pounds of cocoons, of which 427 lbs. 8 oz. belong to the proprietor.

We must also consider that the income increases yearly, if the mulberry-trees are well cultivated; although out of the 1,000 there would only be 750 which can be stripped annually, 250 trees remaining only pruned and fallow.

When the 750 trees are of a size to produce each 30 pounds of leaves, they will yield altogether 22,500 lbs. which may produce 1,605 lbs. of cocoon, the half of which belong to the proprietor.

The calculation I have just offered will be sufficient to shew to those who have the intelligence of business, that there exists no branch of industry which interferes so little with others, and yields greater annual profit, than the cultivation of mulberry-trees and silk-worms.

Besides what I have been saying, there are four circumstances concurring in favour of the proprietor.

1. It is a fact, that however extensive the plantations of mulberry-trees may be, on land already let and tenanted, the proprietor generally makes no compensation or allowance to the tenant for the ground thus occupied.

2. We have supposed the proprietor to buy all his mulberry-plants, and each plant to cost two francs. If he gets his plants from his own nursery, and although he should have the banks dug and manure them thoroughly, yet will not each plant stand him in a franc.

3. The quantity of mulberry-plants that we have averaged would not die, and casualties are exaggerated.

4. If the proprietor go halves with the cultivator in the rearing of silk-worms, he obtains directly, or indirectly, the greatest share of every advantage which the cultivator derives from them, because the produce of the whole of the cocoons ends by almost all coming into the hands of the proprietor.

Besides it being well known that the rent of the farm may be raised as the value increases.

Such is the true state of things. Any who wish to detract from the value of this important art are at liberty to do so. I will only say, that if the proprietor does not attentively watch the plantation of the mulberry-trees, and their culture, particularly during the eight or ten first years, he will not receive from 1000 trees the quantity of leaves which an able agriculturist will get from 200 mulberry-trees at the end of six years.

This observation deserves the strictest attention, because it is the basis of those lasting advantages which may be derived from this valuable branch of industry.

There are further considerations which must be laid before the proprietors.

Let us suppose the 732 mulberry-trees above mentioned, only standards on the land (or even a thousand), besides the 1000 trees also above stated, all planted on a surface of about 2000 feet of good ground. This ground producing three *cartes* of wheat per

perch, would produce 150 setiers, which, at 16 francs
each, would amount to 2,400 francs. The annual
product of the mulberry-trees, or that which in a
few years they will produce, will be equivalent, as
before stated, to 750lbs. of cocoons to the share of
the proprietor.

To obtain these cocoons, the space of land required
will be 2⅕ perches. The value of the cocoons, calculated
at only two francs per pound, will come to 1,500
francs; it is therefore clear that this sum is not only
equivalent to the value of 2½, but to that of 83 perches
of ground, which would each yield three *cartes* of
wheat ; in other words, the proprietor, by adding 2,000
francs to his capital, will obtain a rent equal to that
yielded by 83 perches of good land, which would cost
him infinitely more.

In fixing 200 perches for 1,000 plants of mulberry, I have doubtless fixed ground enough, for in
dry soil of 1,400 feet I have planted above 700 that
are no wise injurious, and I have besides a high mulberry hedge, which produces a great deal.

Having thus concluded all that regards the proprietor, let us now turn to the cultivator or tenant,
who divides the cocoons with him.

I will begin by one fact, which comprises thousands, and which alone can fully display all the others.
In the year 1814, for instance, when most cocoon-crops
failed, one of my tenants gathered 180 lbs. of cocoons
for his share, which was the half of 360 lbs. which
four ounces of eggs produced. This tenant had only
120 perches of land.

These cocoons, which were of an excellent quality, sold for 2 francs 55 centimes per pound ; the 180 lbs. of cocoons, therefore, produced about 456 francs.

The property which this tenant holds and manages is of fine soil.

When the value is compared of the sums which the tenant must advance in expense and labour to obtain the 22 setiers requisite of wheat, the risks he runs before he can reap and sell the corn, when all this is compared to the labourer's expense of labour alone, which is required to produce 360 lbs. of cocoons, the superior advantage of the culture of silk-worms will appear most evident.

The produce of the cocoon is therefore greater to the tenant than the production of any other thing he cultivates.

I must not conceal, however, that besides the price of labour, and the salaries which the tenant pays, he is liable to some losses and further expenses.

1. The mulberry trees cast much shade from the hedge-rows and in the fields, and injure the crops to a certain degree.

2. The ground is much trodden when the leaf is gathered.

3. He consumes fuel, oil, paper, and loses the interest of the capital employed in expenses of hurdles, wickers, tables, and other trifling utensils.

These losses again are balanced,

1. Every year he obtains a great deal of wood from the mulberry-trees, in pruning one quarter of them.

2. He has the dung and litter taken from the wickers while the worms are rearing.

3. And the leaves of the mulberry-tree are used as fodder, when they drop off in the autumn, for oxen.

In concluding this paragraph, I ought to say that the great profits which either proprietor or cultivator can hope to obtain, are founded on the judicious planting and culture of the mulberry-tree, and on the careful rearing and cultivation of the silk-worm; and that these profits will never be reaped by any who do not attend to the plantation of the mulberry, or who do not rear the silk-worm with the most minute attention.

3. *Net profit which may be made by those who rear silk-worms entirely on their own account, and at their own cost, whether using the leaves they have, or buying leaves to feed the worms.*

The calculations I am going to note down, tend to shew that the art of rearing silk-worms may be practised by any one possessing a room, silk-worms' eggs, and mulberry-leaves at their own disposal.

The rules I have laid down are such, that without being either proprietor or tenant cultivator, any one may rear silk-worms with success equal to theirs.

The table of all the expenses I incurred to rear silk-worms, during the two last years, and of what I gained on them, is very exact. My laboratory or establishment would hold the worms proceeding from five ounces of eggs from the fourth moulting or casting, until the complete accomplishment of the changes.

1813. *Expenses incurred.*

	Milan livres.	Sous.
Five ounces of silk-worms' eggs - -	15	0
Fuel-wood for hatching them (Chap. V.) - .	1	15
82 quintals or 80 cwts. of mulberry leaves, at 4 livres 13 sous, average price at 6d. the quintal - - - - - -	385	0
18 quintals 75 lbs. of shavings, thick wood and brushwood, at 1l. 1s. 4d. the quintal	20	0
Boughs or broom to form the hedges for the worms to rise upon - - - -	18	0
Paper to put upon the wicker trays - -	14	0
Oil for the lamps - - - - -	9	0
Fumigating bottle - - - - -	1	10
Expense of gathering the mulberry-leaves, at a rate of 22 sous 8 dec. per quintal -	96	5
100 days men's labour, at 25 sous per day, women at 15 sous; when the men work extra hours at night they get 10 sous more, and the women 5 sous additional, altogether amounting to - - -	103	10
Ground-rent, and interest of capital employed for purchase of trays, wicker, and other trifling articles - - -	90	0
Total -	754 Liv. M.	

Most of the paper and fagots of brooms may serve for the following year.

613 lbs. 8 oz. of cocoons were drawn from the laboratory, which, sold at 34 sous 6 dec., make about - - - -	1063	0
Net profit in favour of the person who has reared the worms on his own account -	309	8 sous.

1814. *Expenses incurred.*

	Milan liv.	Sous.
Five ounces of silk-worms' eggs . . .	15	0
Fuel wood for hatching them . . .	1	15
8250 lbs. of leaves, at 4 livres, 13 sous, and 6 decimes, the 100 pounds or quintal . .	385	0
Expense of gathering the leaves . . .	96	5
Shavings, large and small fire-wood, 15 quintals, at 1 liv. 4 sous the quintal	16	0
Fagots and broom to form the hedge for the worms to rise on, in addition to those of the preceding year, 1813	4	10
Paper, in addition to that of preceding year .	4	0
Oil for lamps	9	0
Fumigating bottle	1	10
For day-labour, men and women . . .	109	0
Total	642	0
Ground-rent and interest of capital employed .	90	0
	732 M. liv.	
601 lbs. 1 oz. of cocoons were drawn from the laboratory, which sold at 52 sous per lb., and produced	1563	18
Net profit	831	18 sous

The result of this calculation is, that the tenant always derives a positive advantage or gain for labour alone: this profit was exceedingly great this year, for those who managed silk-worms well.

By what I have stated, it is easy to perceive that the tenant only furnished a sum of the value of about 200˙ francs in labour, and about 50 francs in money expended, including the price of half the eggs. If this expense is compared with the profit, the result will prove that, in 1814, he made 781 liv. 19 sous, which is the half of 1563 liv. 18 sous, the total amount of the income of the year. It would appear that this is no inconsiderable profit for the tenant; it is earned in a few days, and is certain.

The high price of cocoons in 1814 doubtless contributed considerably to increase the profit. In 1813, cocoons being low, the profit did not exceed 309 liv. 8 sous; however, even this sum ought to be sufficient to encourage the cultivator.

Let us return to the primary object of this paragraph, and conclude by observing, that, from the calculations I have made, the result is, supposing 20 or 21lbs. of leaves yield 1½lbs. of cocoons, the profit is greatly in favour of those who rear and cultivate silk-worms.

That in thirty-five days, during the greatest part of which there is little to do, the cultivator may gain a sufficiency for himself and family to live on for some months.

That a numerous family, bestowing care and assiduity, may, by working themselves, save the expense and price of labour, and, consequently, make more than those who are obliged to pay for all the labour.

That any individual renting a spacious laboratory, and an abundance of mulberry-trees, may, in a very few years, upon the sole value of the cocoons, obtain an income equal to that of a tolerably good property, I say tolerably good, because I must always be understood to speak about small proprietors, who are by need compelled thoroughly to cultivate their own pro-

R

perty. Large proprietors, in general, do not attend much to the augmentation or amelioration of the production of their possessions.

4. *Annual Increase of Riches which the first general Improvement in the Cultivation of Silk-worms would introduce into the Nation.*

A country may increase in riches, as a private family augments its wealth, in two ways.

The first, in raising the value of annual production, without increasing annual consumption.

The second, by diminishing annually the ordinary consumption, if there is any difficulty in raising the value of annual production.

The first object is obtained by improving national industry, and the second is attained by economizing or diminishing useless consumption.

These two modes of augmenting riches shew that the increase and decrease of national or private wealth depend on the individuals that compose it, and not on the nature of the government. It is only the acts of administration that have any direct influence on the augmentation and diminution of produce, and annual value of such produce, inasmuch as they can encourage or check sale, and consequently encourage or check the zeal and labour which those who give work or sell would naturally employ.

I think I can demonstrate that Italy can produce in a few years above forty millions of exportable silk.

And in evidence of this truth I will here place the following statement. At this time the quantity of raw, spun, and dyed silk exported (leaving out the rest) abroad, proceeding from the provinces which composed the ex-kingdom of Italy, amount to eighty millions, an amount less than in 1810, of which the following is the account, (§ 1.) :—

Livres:

1. I will suppose that 90 lbs. of cocoons are obtained from an ounce of eggs, instead of 45 lbs., which, on an average, is now the common quantity. Half the number of cocoons kept for the produce of eggs will be saved. (Chap. IX.) This half of the quantity of cocoons, which would yield exportable silk, amounts to . . . 800,000

2. If silk-worms are reared so that 21 lbs. of leaves should produce about 1½ of cocoons, whilst just now 25 lbs. of leaves are wastefully consumed to produce that weight, as, on an average, they reckon 1050 lbs. of leaves necessary to produce 60 lbs. of cocoons; we should thus have, undeniably, a quantity equivalent to a quarter more; its value would then rise to about . . . 20,000,000

3. I will suppose that by good management the losses decrease only 10 per cent., this will produce in more or less time 8,000,000

4. Admitting that the cultivation of the mulberry-trees improves and plantations increase, we must suppose the production of leaves will augment 10 per cent., which will be an increase of the production of cocoons amounting to . . 8,000,000

5. Supposing the amelioration in the art of rearing the silk-worm, the cocoons obtained from good and well-managed laboratories, in equal quantities with those obtained from ill-directed laboratories, will yield 10 per cent. more silk, as experience every day testifies. The value of this amelioration will amount to 6,000,000

Total 42,800,000*

* Count Dandolo had been employed for many years in making experiments on the mulberry-trees, when he died by an

This calculation may appear exorbitant; but on reflection, I am fain to believe most persons will find it
moderate.

I was obliged in this account to diminish the value
of exported silk, confounding it with the production of
the cocoons; I could not do otherwise, as silk is exported, and not the cocoons.

From all I have said and demonstrated, there is a
certainty, that in augmenting the annual production
and exportation of silk by forty millions only, there
would be a net profit to the country of two-thirds, or
above twenty-six millions a year.

Hitherto I have only considered silk as an exportable production; but it must evidently appear that its
annual value augments prodigiously, when we reflect
on the quantity which is externally required for our
wants, and by our habits.

It is certainly not easy to foresee to what sum the
value of exportable silk may rise, if the art of cultivating it becomes national, and the objects of the care
and attention of intelligent, scientific, and patriotic

apoplectic stroke in his villa at *Varese*. He was lamented by
the people, to whom he was a father, and will be long regretted
by those who know how to appreciate distinguished and useful
men in domestic economy.

He had told me, and his brother-in-law Mr. Grossi has confirmed me in the idea since, that he did not hope to obtain
much more satisfactory results, on the culture of mulberry-trees,
than those which had been published some years since by his
friend Count Verri.

I have stated in my advertisement that I hope soon to publish
this book in French, which, being a short volume, will be
cheap; I shall add to it a few notes found in M. Dandolo's
manuscripts communicated to me by Doctor Grossi.—(*Translator.*)

individuals. Hitherto this most valuable pursuit has only presented a mass of various and different methods, most of which were uncertain, and many absurd.

The manner of manufacturing silk may vary in different civilized countries, according as the fashions change, but silk will never cease to be most eagerly sought after among all nations. No natural or artificial production can vie with silk either in magnificence or brilliancy. Courts and nobles may in vain seek, in any other material, ornaments to gratify their vanity or their luxury; and the temples of religion could find nothing more sumptuous to decorate their high solemnities.

The object, therefore, is to produce silk in quantities either raw, spun, or manufactured, to supply the whole globe with it.

The various reckonings and calculations that I have made were always grounded upon authentic facts drawn from the registers of the administration of the ex-kingdom of Italy. Each Italian province, that was not included in this kingdom, might calculate the amount of exported silk within itself, and in this manner might be known the immense value of the silk the whole of Italy exports.

Fortunate shall I be, if, having examined the detail of this great object of national industry, I may, by inspiring the wish of rearing silk-worms skilfully, contribute to ameliorate the condition of the industrious.

THE END.

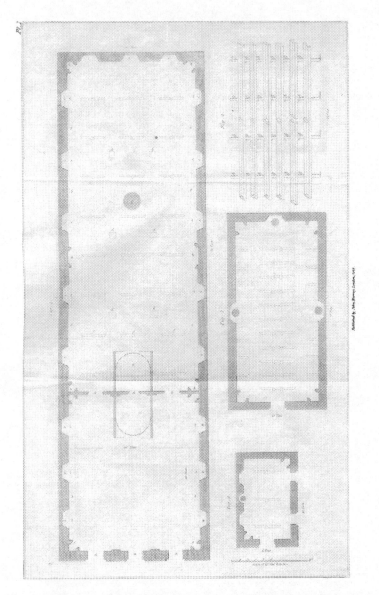

The material originally positioned here is too large for reproduction in this reissue. A PDF can be downloaded from the web address given on page iv of this book, by clicking on 'Resources Available'.

EXPLANATION OF THE PLATES.

PLATE I.

Fig. 1. Great Laboratory, with the Hall or anterior Apartment.

a. Six doors; three of which lead into the hall, and three into the great laboratory.

b. Six windows, having each a ventilator placed on a level with the floor, opening at will.

c. Small room or passage, in which is a trap or semicircular opening, by which the dung is cast out, and the leaves brought in with a pulley.

d. Six ventilators in the floor of the great laboratory, to facilitate the circulation of air.

e. Windows, under which ventilators are placed, in the hall or anterior apartment.

f. Figures, giving an idea of the regular disposition and manner of placing the hurdles, wickers, or tables, in three lines or rows.

g. Stove.

h. Six grates or fire-places :—

1. Large stoves for drying the leaves in this apartment.

2. Valves with a small cord in the centre, and two knots keeping the valve open; the cord is fastened to a large nail with a knob head, fixed in the middle of the window-seat.

3. Sign showing the spot where the thermometer and lamps must be fixed; the former are fixed in the lower part of the wall, between the windows, and the latter in the upper part of the wall. ∫

LIST OF PLATES.

Fig. 2. Laboratory yielding about 600 lbs. of cocoons ; it contains four fire-places in the angles, a stove in the centre, and another opposite the door.

Fig. 3. Small laboratory, with two fire-places in two angles, and one stove.

Fig. 4. It may be seen that the tables or wicker trays are placed one after the other, so that several which are 15 or 18 feet long, may, by being put together, form a single table of 55 or 75 feet long, and upwards.

PLATE II.

Fig. 10. How the ventilators, which are placed in various parts of the laboratory, open and shut at will.

Fig. 13. Travelling case for removing silk-worms any distance ; each board draws out easily and slides in, and is large enough to hold a sheet of paper containing an ounce of eggs. If there are only a few ounces to be conveyed, the supernumerary boards may be taken out. The case is carried like a pedlar's pack, on the back.

Fig. 23. Fumigating bottle. In turning the screw, the stopper is raised; which stopper is made of ground glass fitted to the bottle, shutting it hermetically ; the neck of the bottle should be ground also, as the two ground surfaces fit more closely. When it is opened, it may be removed from its stand, and carried any where.

All the other Figures are objects easily understood and distinguished, and need no description. (Chap. XIII. § 4.)

LONDON
Printed by W. CLOWES, Northumberland-court.

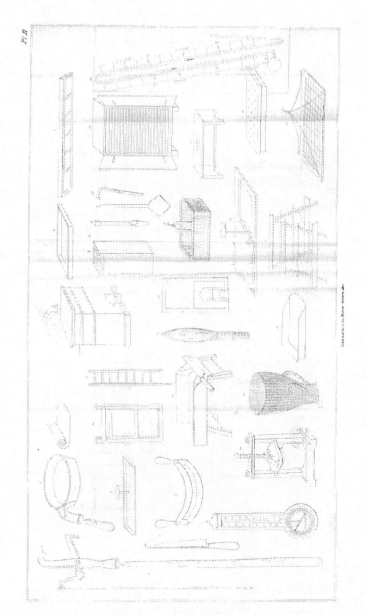

The material originally positioned here is too large for reproduction in this reissue. A PDF can be downloaded from the web address given on page iv of this book, by clicking on 'Resources Available'.

Printed in the United States
By Bookmasters